MOVING BRIDGES

MOVING BRIDGES

BRIDGES

OPERATING THROUGH CORRUPTION
TO JUSTICE AND FREEDOM,
AND A CALL FOR NEURORIGHTS

DIANA JANDRESKI

ISBN: 979-8-9985123-0-8 (pbk)
ISBN: 979-8-9985123-1-5 (hbk)
ISBN: 979-8-9985123-2-2 (ebook)

In hopes of a safer and healthier future for all.

TABLE OF CONTENTS

PREFACE

This book provides a detailed account of my experiences working at HDR Inc. as a Movable Bridge Engineer. As a large corporation in the engineering and architecture industry, HDR has a global presence and a workforce exceeding 10,000 employees. For a considerable period, I envisioned a long and prosperous career within this company. Unfortunately, after approximately three and a half years, I found myself at a crossroads, faced with some profoundly challenging decisions.

Despite the adversity, my tenure at HDR proved to be an invaluable learning experience. As a result of this time, I gained insights into handling workplace misconduct, understanding employment law, and how to protect yourself when faced with corruption and bad faith.

While navigating the path to recovery, I have been compelled to share my story and the lessons learned along the way. My silence would only lead to regret, knowing that my experiences could potentially aid others in similar situations. This book aims to shed light on the corrosive and destructive impact of corrupt corporate power, serving as a cautionary tale that may help others learn from my own difficult experiences. It is, above all, a tool for raising awareness.

In these pages, I delve into a range of experiences, including sexual and psychological harassment, discrimination, retaliation, and whistleblowing on fraudulent activity before eventually being forced to resign. I detail the subsequent negligence and bad faith actions with continued retaliation, both in the workplace and outside of it. From manipulating circumstances and conspiring to create false narratives to misconduct by law enforcement and medical malpractice. My commitment to telling the truth is unwavering, for I believe in accountability.

In addition to raising awareness on current laws and the justice system, and arguably most importantly, this book is meant to provide a call to action for necessary regulations and laws regarding mind data privacy, artificial intelligence (AI), and neurotechnology. With this type of unchecked technology, corruption and malice can cause painful impacts on individuals' lives—which could be preventable with transparency. Sharing my story is not just cathartic; it is a vital step towards instigating change and offering support to others grappling with similar challenges.

1

INTRODUCTIONS

INTRODUCTION TO MOVABLE BRIDGES

You may be wondering, "What's a movable bridge?" As you're presumably aware, a bridge is a structure designed to span an obstruction or obstacle, such as a river. A movable bridge is specifically used over navigable waterways where a fixed bridge is either not feasible or undesirable. It operates to open and close as necessary, allowing for the passage of large boats, ships, and barges below when open, while still providing a way for vehicles, trains, and pedestrians when closed.

There are three primary styles of movable bridges: vertical lift, swing span, and bascule style bridges. A vertical lift bridge translates vertically up and down along tall towers situated on either side of the bridge, facilitated by a machinery system composed of ropes and sheaves or pulleys. The swing span bridge operates by swinging open and closed on a vertical axis, typically pivoting at the bridge's center. Lastly, the bascule style bridge—named after the French word for 'seesaw'—rotates back to open in a manner similar to the children's playground equipment after which it's named. The longer leaf or span section of the bascule bridge on one side is balanced on the opposite side of the pivoting horizontal axis or fulcrum by a large counterweight. For the same reasons, when on a seesaw, leaning backwards or forwards can make you rise or fall, a counterweight ensures an efficient machinery system for opening and closing the bridge.

Movable bridges require the collaboration of three main engineering disciplines: structural, mechanical, and electrical engineering. These professionals work as a team to provide the necessary expertise for movable bridges. Structural engineers focus on the bridge itself, mechanical engineers handle the machinery to move the bridge, and electrical engineers manage the power and control systems required for operation.

When I refer to "the bridge" that structural engineers are responsible for, I am talking about the physical structure that mainly bears its own self-weight and the loads imposed by traffic. This includes the deck, which is the top layer that vehicles, pedestrians, and other users' interface with—it's what you are driving on when in your vehicle. The deck transfers the traffic loads through the superstructure to the substructure. The superstructure could range from a simple system of I-beams to something more complex like a truss, arch, or suspension system. Regardless of the type, the deck and superstructure are supported by the substructure. Substructure elements typically consist of large concrete masses that absorb the high compressive loads from above and transfer them down to the earth.

From the mechanical engineering perspective, there are generally two main types of drive machinery systems designed and constructed to move these large bridge structures: electro-mechanical or hydraulic systems. These systems usually generate rotational motion to drive the bridge, similar to how the rotation of your car's wheels propels it forward. Electro-mechanical systems incorporate main motors and brakes, as well as a drive train that could include open gearing, enclosed gearboxes, shafts, bearings, and couplings. Hydraulic systems, conversely, are powered by a Hydraulic Power Unit (HPU) comprising of electric motors, hydraulic pumps, valves, hoses, and piping. The HPU delivers the pressure and flow required to open and close the bridge with hydraulic fluid. This fluid is pumped through hoses and piping, typically to hydraulic cylinders, which provide the linear motion required to drive the bridge. In addition to the main drive machinery, there are also ancillary mechanical systems, such as span lock systems.

Electrical engineers play a crucial role in providing the necessary electrical power and controls for bridge operation. Electrical power is supplied to the bridge by a service provider, with transformers used to drop the voltage as needed. As for the controls, they are usually housed in a control house manned by a bridge tender or operator, ready to respond to a call to operate the bridge. Some bridge owners have transitioned to remote operation. Most modern movable bridges feature either automated or manual push-button controls centralized at a control desk. These controls are designed utilizing sensors to ensure the bridge operates in a precise sequence and timing, ensuring safety and smooth operation.

In terms of the types of projects that movable bridge engineers work on, there are again generally three main categories: new design contracts, rehabilitation design contracts, and inspection contracts. These job types vary significantly in terms of their requirements and deliverables.

A new design contract involves constructing a completely new movable bridge. This might be required at a brand-new site location, but more commonly, it involves replacing an existing bridge that is nearing the end of its lifespan. Alternatively, a bridge owner might want to replace an existing bridge because it no longer meets the necessary travel requirements; for example, updating a bridge from two lanes to four to alleviate traffic congestion. A new design project requires a considerable amount of manpower, detailed work, and coordination among the various engineering disciplines.

My personal favorite is the rehabilitation contract, as I have a passion for fixing things. In these projects, engineers are tasked with inspecting and assessing the current condition of an existing bridge. Based on these evaluations, recommendations are provided to the owner on which components need repairing or replacing. Occasionally, multiple alternative recommendations are provided, and budgetary constraints may influence the final decision.

Finally, inspection contracts involve either routine or in-depth inspections intended to monitor the bridge's condition, recommend maintenance work, and maintain a historical record of the bridge's state. This is akin to having an annual check-up with your doctor. During these inspections, teams visit the site to take notes and photographs of the various bridge elements, comparing them to previous inspections. The collected data is then compiled into a report for the client's records.

Still, you might be wondering what engineers are actually doing on a day-to-day basis. Besides going to the field to inspect, assess, and report on the condition of a bridge, movable bridge engineers also do a ton of office work related to design. When completing a new or rehabilitation design, movable bridge engineers perform necessary calculations to ensure the designs meet the required standards for functionality and reliability, adhering to movable bridge design guidelines. This means verifying that each component can perform its intended function consistently throughout its expected lifecycle. The office work also involves using 2D or 3D modeling software to create detailed drawings and plans (comparably some industries call 'blueprints'). Contractors later use these plans along with written specifications from the design contracts to carry out the construction work. While engineers develop the designs, it is the contractors who bring them to life and build it.

INTRODUCTION TO ME

My interest in engineering has been obvious since childhood. I was always curious about how things worked and found joy in fixing them. Math and sciences came naturally to me. Although English is my second language, here I am writing a book out of necessity to pull myself out of a dire situation. If my words can help others and raise awareness to drive positive change, then it's worth the effort.

A great source of strength and fortitude within me comes from my family. My parents legally immigrated from North Macedonia, a country that was part of former Yugoslavia. Located in southeastern Europe within the Mediterranean region, North Macedonia is a Slavic country with stunning landscapes, rich culture, and delectable cuisine. I am proud of my heritage and sometimes cook Macedonian dishes to bring a taste of home into my life. It is a big part of who I am and how I was raised.

I grew up in a suburban town in New Jersey, outside of New York City. There was a sizable Macedonian community in the area, centered around a few churches not far from my home-town of the Macedonian Orthodox Christian denomination. My family wasn't extremely religious; we didn't attend church every week, but the concept of God was instilled in me from a young age. My father would often say things like "God knows" or "God's watching," which encouraged me to be honest and strive to do the right thing, something for which I am grateful.

About 20 years ago during a visit to North Macedonia as a young adult I gained a new perspective on my background. For the first time in my life, I understood the environment and circumstances that shaped my parents. I could see where they came from and how they lived during their youth, which fostered a deep sense of appreciation. More than ever, I now see that life is a series of events and circumstances that carve a path. All of life's experiences are necessary and teach you things along the way.

This trip also opened my eyes to the beauty of my fami-ly's homeland. During my trip, I visited major cities like Skopje,

Prilep, and Bitola, each rich in history and distinct in their own way. Ohrid was another city I explored, which quickly jumped to the top of the list. At the time, I didn't know that Ohrid is globally recognized for both its natural beauty and wildlife, as well as its cultural heritage. I have always admired nature and wildlife. To me, Ohrid most stood out for its lake shore-line, a notable feature since North Macedonia is a landlocked country. I have a liking for being in and around water, finding it immensely peaceful.

This trip took place during the summer when I was nineteen years old. I had been accepted to Rutgers University right out of high school, but I didn't end up attending. Instead, I enrolled in community college that fall and eventually in my engineering school of choice. What seemed like a setback at the time, now seems like a course correction, leading me down the path towards movable bridges and even my recent forbidding experiences.

Actually, it was during my high school years, while taking an auto shop class as an elective, that I first heard the term "engi-neer"—or at least that was the first time it truly registered with me. I took auto shop because it seemed interesting and fun. We covered some basics, such as changing a flat tire and the engine oil. However, two main projects stood out to me. One involved disassembling and reassembling a drum brake, timed as if we were soldiers in the military assembling rifles. I take pride in saying that I achieved the best time in the class. For our group project, we had the opportunity to dismantle an engine and then rebuild it. The requirement for passing the project was ensuring the engine could turn afterwards. Needless to say, our engine turned when we were done.

During my time at community college, I had made up my mind: I wanted to go into engineering. I also worked as a tutor at the community college, assisting students in a variety of classes ranging from basic math to precalculus, physics, basic logic, and an art class. It was a rewarding experience, and I am grateful to have been able to work with each and every one of those students.

Later, I applied to Stevens Institute of Technology and was accepted. With most of my general electives and humanities courses already completed, I was thrilled to begin studying engineering. I continued working through school and started a position as a grader for one of my preferred classes. The class was called Mechanics of Solids, and it covered the basic principles of both mechanical and structural engineering—a perfect fit for me. Despite any challenges I have faced in life, my strong work ethic, cultivated throughout, has helped me through.

I have a bachelor's degree in mechanical engineering with a minor in structural engineering, and a master's degree in civil engineering, concentrated in the field of Structures. Sounds like a lot. How that all happened is quite simple: I was initially drawn to engineering by the mechanical discipline, but while in school, I decided I was not keen on the expected career options for a mechanical engineer. It's not just about the material you are learning, you should also seriously consider what type of work you will be doing when choosing a career. The three main industries for mechanical engineers seemed to be in defense weapons, automotive, and heating, ventilation, and air conditioning (HVAC). I couldn't face the idea of being responsible for working on weapons and HVAC did not seem like something I would find interesting in the long term—no offense to those who work in HVAC or weapons. Finally, while I might have been interested in automotive, it seemed to involve relocating to places where I didn't want to live.

Therefore, I asked myself, "What would be the next best option for me?" The answer was bridges. But, unwilling to give up on mechanical, I kept that as my major and added the structural. When entering industry, I worked on fixed, or non-movable, bridges for a while, gaining experience in that area. Early in my career, I noticed there was a niche within bridge engineering: movable bridges. I realized this was a seamless fit and from then on, my career path was clear. I moved from New Jersey to Florida, making the leap and transitioning from fixed bridges to movable bridges. In the latter part of my career, I mainly worked within the mechanical discipline of movable bridges, operating out of Florida. I had the privilege of traveling around the state, country, and even Canada for work. It has been full of amazing experiences and I will always have a fondness for movable bridges.

INTRODUCTION TO THE FLORIDA MOVABLE BRIDGE GROUP

I started at HDR in the summer of 2019. I was hired as a mechanical engineer in the movable bridge group out of the Tampa, Florida office. I had recently passed my licensure exam, so I was officially a Professional Engineer! I don't think I will ever be able to paint a picture clear enough for you to understand how excited and proud I was of this achievement. I was in a good place in my life with a real potential to grow and succeed to be what I had been striving towards for most of my life. I was also eager because I had known of HDR through the industry, and I thought it seemed like a great place to work. Ready and hopeful, I looked forward to a long career with HDR. Little did I know what was coming.

At first, everything seemed fine. Besides myself, the Tampa group at this time consisted of two other mechanical engineers and two electrical engineers (not counting management). It was a fun group, and everyone had their own distinct personalities.

Randall was one of the mechanical engineers. He was tall and thin, and he knew how to deflect conflict with humor. A good member to have on the team in many ways. Randall had more years of experience than I did, but did not have his engineering license. In fact, none of the other engineers in the group had their licenses. Randall didn't seem to be at all interested in moving up in the company and later gave his resignation notice to become self-employed after buying some real estate.

The other mechanical engineer on the team was Thomas. He was an average-looking guy but quite a character. If he was in the room, you would probably know it. He liked to talk, as well as complain, and often be inappropriate. Sometimes he would say things, giving me pause and I would try to bring the conversation back to work. He was the least experienced out of the whole group, but he was also the most schmoozy or brown noising. From the beginning, it seemed like he didn't want me there. I figured that would eventually go away as we got to work together.

Then there were the two electrical engineers, Duane and Marcus. Duane was more senior and the controls specialist on the team. He had somewhat more of a flashy style to him: he wore glasses, had a fancy watch, and dressed in skinny jeans and dress shoes with colorful socks. Duane was also very personable and would usually be the one to start small talk or crack a quick joke. He was very direct, which I appreciated, since I can be direct myself.

Last but not least, Marcus was our electrical engineer with a primary focus on power. I'm not entirely sure what to say about Marcus. He was smart, but that was a trait shared by the rest of the group. He tended to keep to himself, concentrating on his work, and then swiftly transitioning to his personal life—a sort of 'get in, get out' mentality.

Meanwhile, my life over the years had become balanced around work. Overall, I kept to myself and thoroughly enjoyed working. Life had propelled me further into education and then into my career. I can enjoy a good laugh at a fun joke and engage in chit-chat but I also kept things professional. Typically, I worked on mechanical tasks, but with my background, I was sometimes assigned structural tasks or even asked to review work for electrical. Genuinely, I was always happy to assist. I believe it is part of my culture and background to want to serve. But for me being career-focused was also a personal choice. Being somewhat quiet and having a strong work ethic isn't against the law, but discrimination and fraud certainly are.

So here I was, a female mechanical engineer, working in an extremely male-dominated industry. This was our group before getting into management roles. It was a solid team to be a part of—well-rounded and appropriately suited for working on movable bridges.

INTRODUCTION TO BAD FAITH

In simple terms, bad faith involves the intent to deceive or lie. It represents a lack of honesty in interactions with others. While it is universally regarded as wrong, enforcing honesty is a complex issue. How can you prove the intention to lie when someone could easily lie again? Upon reflection, the objective to tell the truth isn't always simple. Some people lie with malicious intent, aiming to deceive for personal gain. However, others might lie to spare someone's feelings, such as in white lies. The intentions behind the action play the most significant role in defining the nature of the lie.

Are white lies okay? When someone tells a white lie, they are lying; however, their intentions are good. Does that make it wrong? Perhaps you give a compliment on an unflattering haircut, or tell someone their new outfit looks great even though you don't like the style. A coworker gives a presentation for the first time and bombs—do you tell them it was horrible, or do you say, "Good job"?

Being brutal with your candor and telling a coworker their presentation was terrible can discourage them, potentially leading to even worse performances in the future. So, is there a way to be both truthful and helpful? Absolutely. Providing direct but compassionate feedback, highlighting positives as well as areas for improvement, proves most beneficial for everyone involved. So then, is a white lie with good intentions necessarily bad?

This brings us to the topic of ethics. What is the ethical thing to do? Some might argue that whatever benefits the greater good is the right choice. If complimenting someone's haircut boosts their self-esteem, isn't that contributing to the greater good? After all, it's not harming anyone. However, others could counter that people should be resilient enough to handle the truth, even when it's negative. By not offering your honest opinion, you might be perceived as coddling them, shielding them from reality. However, your opinion is only your perspective. Nevertheless, honesty can be delivered compassionately and with respect.

Now, consider lies explicitly meant to harm others, lies told with the intention of causing damage. That is unequivocally wrong, manifesting as manipulative and malicious behavior. Lies told to gain an unfair advantage, cover up unlawful misconduct, or commit fraud are undeniably reprehensible. Deception that leads to neglecting one's duties or responsibilities. These examples can all fall into the category of bad faith. But how often does this occur without consequence or accountability? And how frequently do people engage in bad faith, creating a snowball effect that results in a huge and intricate web of lies? As you continue reading, you will encounter instances of bad faith that continue to spread in just this manner.

2

MY DIRECT SEXIST SUPERVISOR

MY BOSS, NICK

What exactly is a supervisor? Is it someone who oversees your work and assigns tasks, provides leadership, and aids in your professional development? Or someone who potentially discriminates against you and makes your work life unbearable? Could it be someone with the power to ruin your career? All of these examples are possible. What you hope for in a supervisor is someone who acts as a leader providing support and not as a boss that causes you harm.

A boss is someone who dictates, instills fear, and hinders your growth. A leader, on the other hand, is someone genuinely interested in assisting and fostering your development and career progression. The nature of your supervisor can be the deciding factor in the trajectory of your career and overall life.

During my time at HDR, my direct supervisor was unfortunately a harassing boss rather than a supportive leader. His name was Nicholas Stone, though he went by Nick. Nick was of average height and build, typically dressed in jeans and a polo, and sported a short, buzzed, military-style hair-cut. Like anyone, he had his good and bad days, but his bad days tended to negatively impact those around him. Nick was mean and cruel—or "harsh" and "gruff," as some would later describe it in attempts to downplay my complaints and his behavior. He employed intimidation tactics to drive with fear.

Then there was me, a female mechanical engineer, work-ing directly under him since he held the positions of lead mechanical engineer for our group as well as group manager. He was knowledgeable technically with regards to mechani-cal engineering and movable bridges, and could have been a great resource as a mentor. However, Nick did not see me as the most qualified engineer on his team, nor the most diligent and hardest working, regardless of my performance. In his eyes, I was merely the woman of the team, and that was how he treated me. Nick was sexist, and he repeatedly harassed me and discriminated against me.

Apart from the contrasting styles of a boss versus a leader, which have been increasingly discussed in workplace contexts, another term for a person in a position of power is 'tyrant'. Do I consider Nick a tyrant? I believe this term should not be used lightly. A tyrant is someone at the pinnacle of power, exhibiting no consideration for others and oppressive cruelty to all, irrespective of bias.

Nick was not a tyrant. He was the group manager, but he had supervisors himself. He could at times be mean to any-one; however, his behavior was typically directed. He was usu-ally supportive of Duane and Marcus. When it came to the mechanical engineers, though, he was less restrained. And although at times he could lose his temper with Randall or Thomas, it was nothing like the harassing, aggressive, degrading, and sexist treatment that I endured from him.

It wasn't long after I started working there that it all began. At first, he seemed nice, and we would have discussions about work and my tasks, as expected from a supervisor. But then an underlying tone and slight nuances started to surface after only a few weeks, along with insinuations and implications. The sexual harassment started after about two months. I didn't speak up or talk about the sexual aspects of his behaviors for a very long time—not until I was forced to and began to understand my rights much later.

LACK OF PROFESSIONALISM AND FIRST MISTAKES

Now, years after the start of my employment at HDR, I can look back at it all and see where I could have done things differently. In life people go through things that knock them down, but you can always overcome them, get back up, and kept going, even wiser and stronger. The big picture here is much larger than I can sum up, so for now let's start at the beginning.

I didn't start documenting things until about a year and a half after my employment started. But it was only around a month after I started that Nick first flipped his lid on me. I was at my desk working, and he came over to discuss something. I don't know if I asked him to come over, or if he just dropped by, which he would do now and again. Our office was partly open plan, with cubicle workstations, and only higher-up management had offices. We started talking about the task I was working on, and I showed him part of my incomplete work. He saw something that he didn't like, and he just went off. He was yelling at me and started putting me down. This was in the office, mind you, with several people around, so it was very embarrassing. I didn't know how to react. Eventually he stopped and walked away.

The next day, Nick called me into an office to talk about what happened. Even Nick didn't have an office at this time and worked out of a cubicle. With some privacy in the room we were using, he actually apologized for the way he acted. He knew it was wrong. I could see it not only through the words that he was saying but also through his actions and demeanor.

Along with being quiet, I could also be agreeable. This was my first mistake and something that you as a reader should not take lightly. My mistake was to accept his apology in a way that was too agreeable. I was being empathetic and concerned about his feelings. From my perspective, I was being kind and trying to build trust in our professional relationship in a positive way. I believe Nick, though, took this as a sign of weakness and me being someone he could take advantage of. His mistake was confusing kindness with weakness.

Although my intentions were good, looking back I know I could have emphasized more that his actions were unacceptable. I could have spoken up about how embarrassed it made me feel, maybe even said that it's not right to speak to one's employees like that. I should have at the very least kept a stern face while accepting the apology. From here on out things got increasingly worse as he kept testing limits and overstepping professional boundaries.

SEXUAL HARASSMENT

For a long time, this section made me uncomfortable. For me, sexuality is something that should be kept private and for serious relationships. Also, what I have realized through this experience is that sexual harassment adds an element of shame that can prevent the victim from addressing it.

It was now about a month after Nick apologized about his previous behavior and about two months since I started working for HDR. I was working late one evening, and this particular evening Nick decided to stay late as well. There was no one else around, so I guess he decided to make a move. Our desks were within the same area, although he was still approximately 20 feet away. He stood up and called me over to look at something. When I walked over, he asked me to sit down in his chair. He said he wanted me to review what he was working on. I sat down and started looking at the computer monitors when, all of a sudden, he slid his arm down the back of the chair. He rested his hand on the seat of the chair right next to my buttocks. I didn't say anything and froze. I should have gotten up and said something. I should have said something such as, "what do you think you are doing?", "stop that", or "no!"

An uncomfortable and scared feeling took over though. He knew it, and after he paused for a few seconds waiting for a reaction from me, he eventually pulled his arm away. I told him I had no comments on the work and rushed out of there. This ought to have been my first red flag as to how sexist he was. Afterwards, I never talked about it to anyone until over three years later. I should have talked to someone about it and reported it back then.

By some means, we believe that if we speak up about this sort of thing the situation will get worse. While that may or may not be the result of speaking up, staying quiet is far worse. Staying quiet meant he got away with what he did. It also meant that he could get away with more. Maybe I thought it would stop on its own or that I could somehow make it stop without things getting worse. Neither of those things happened.

To help in preventing sexual harassment we need to get better at breaking the stigma associated to it and the misconceptions around speaking up about it. With that I list several trigger words for sexual harassment that could have helped during my complaints: inappropriate, stare, ogle, touch, caress, rub, push, thrust, buttocks, breast, groin, penis. If it makes you uncomfortable, I understand and that is the point. Saying these words aloud or visualizing using them while reporting harassment can help reduce the sensitivity around these words. It also seemed like if I spoke these types of words, I would be labeled unprofessional which is wrong. Nick was the one being unprofessional. This was sexual harassment, which is a type of discrimination, and this was not the only time he would do it.

As I mentioned earlier, Thomas would often be inappropriate as well. Nick saw Thomas as a "man" much like himself. In attempts to back Nick up, there were times when Thomas would also make gestures or insinuation to try to push me to have sex with Nick. Thomas told me that he had a small penis but Nick had a big penis. He also would make phallic gestures with his arm. Again, I would ignore it, try to pretend it wasn't happening, and hope it would go away.

Misconduct outside of the limits of the office can also be considered harassment within the workplace. While away for work on bridge inspections, we would often have to stay in hotel rooms. Even here, Nick would try to lure me into his hotel room and I would have to find excuses to get away. It made me feel not only disgusted but stuck. The implications to and worries about my job security and threat of my income were very concerning.

Nick would find ways to get me alone and give me prolonged, inappropriate hugs. The hugs involved him pulling me in and caressing my back and arms, while I just wanted to get away. This always made me feel very uncomfortable but the most I could do was muster an uneasy and embarrassed chuckle, followed by something like, "I gotta get going." It was clear I didn't like what was happening, and yet he continued it anyway. These are more moments when I wish I had said, "Please stop, this is making me uncomfortable." It's not always easy to say something like that though, especially to your boss.

Another thing he repeatedly did over time occurred in the office. I have already mentioned how he would drop by my desk. Well, sometimes while doing this, he would deliberately press his groin area against the back of my chair while I was sitting in it. I would freeze, as though I couldn't say anything even if I tried.

There was one particular instance that was even more heinous and objectifying. While at my desk, he paused and looked around to see if anyone was watching, then he actually thrusted his groin against the back of my chair and laughed about it. I fell forward, bracing myself against my desk. This took me even longer to admit, but his actions here constitute sexual assault and is an even greater offense. Although there may not have been someone who directly saw what Nick did, I was left with a feeling of "what if" someone did recognize what had happened, an unsettling feeling of shame.

By this point, he must have realized I was too scared to say anything. At the same time, the more extreme the harassment became, the more scared I got. On top of having to deal with the sexual misconduct, I also had to contend with the added aggression I received from him when I disregarded and rejected his sexual advances. These instances were when his aggression peaked. It all interfered with my work and assigned tasks, causing me stress. I think shame and fear of Nick, as well as apprehension about what would happen if I did speak up, were mainly what held me back.

However, managers, like Nick, should be trained on this type of issue. They should not only be trained on what constitutes sexual harassment, but also on how to read people and gauge if they are offended or uncomfortable with something. Based on Nick's actions and behaviors, he knew and he continued anyway.

PSYCHOLOGICAL HARASSMENT

Harassment does not have to be sexual in nature. It can be based on gender, sex, race, skin color, age, national origin, religion, disability, and a few other categories. These categories are referred to as "protected classes." Harassment is considered any unwelcome or offensive conduct currently based on one of these protected classes. There are two scenarios in which harassment becomes unlawful.

The first scenario, as it pertains to employment, occurs when the harassment becomes a condition of your employment. This is easier to understand in terms of sexual harassment; for instance, if someone said to you, "Have sex with me, or you'll be fired." This is referred to as a "quid pro quo." However, there are other ways besides sexual favors that this can occur. What if you complain about your boss's offensive and threatening behavior while requesting a new supervisor? If they refuse to accommodate your request or address the problem, aren't they essentially making the unwelcome behavior a condition of your continued employment?

The second way unwelcome conduct becomes unlawful is when the behavior is "severe or pervasive" enough to create working conditions that are intimidating and abusive, called hostile work environments. So, what does that mean? The conduct or behavior can be verbal or nonverbal, encompassing things such as offensive jokes, slurs, name-calling, intimidation, put-downs, and the display of offensive objects or photos. "Severe" refers to the harmfulness of the harasser's actions or words. "Pervasive," on the other hand, relates to how frequently the incidents occur. However, it's important to note that even a single occurrence can be enough to constitute harassment.

Referring back to Thomas's behavior, which was also sexual harassment, by the time I started complaining Thomas was no longer working for HDR. He had given his resignation. Later, he did return to work for our group but Nick kept us separated on task for the most part. In this case, Thomas's behavior was not pervasive, although possibly severe. Nick's behavior on the other hand was both severe and pervasive.

Since this discussion revolves around the workplace, the question becomes: Does the conduct make it more difficult for the victim to do their job? However, the overarching question is whether the conduct is causing harm to the victim. Is the behavior damaging to the individual? Is the offender being oppressive? To be oppressive is to unjustly inflict hardship and constraint, typically on a minority or subordinate group or individual. Such behavior can become burdensome, causing discomfort, anxiety, or depression.

I mentioned at the beginning of this chapter that Nick was not a tyrant. A tyrant is someone who, when they walk into a room, causes every person in that room to feel fear because with a tyrant, no one is safe. Logically, while a tyrant is oppressive, an oppressor is not necessarily tyrannical. Nick was oppressive towards me. His repeated actions throughout my employment caused me a great deal of pain and suffering. If we focus on the work aspect, he acted like a barricade, intentionally hindering the development of my career.

Not only that, but his actions, as well as the dismissal of my complaints about his behavior, caused immense stress. The psychological harassment created feelings of fear and anger which I was much more able to speak up about than with the sexual harassment due to the component of shame.

Nick's issue with me was that he couldn't see past the fact that I was a woman. He was offensive not only towards me, but also towards the female gender as a whole. He harbored a prejudice against women, which he demonstrated in various ways. On one occasion, he made jokes about his wife needing to visit the hospital several times in a row, saying, "The doctors think I'm beating her." I found this far from amusing although he laughed at it. Another time, during a group outing, a waitress approached from Nick's side, and he jolted his arm back as if to elbow her in the face, and again laughed about it afterward.

Many other times though, he also exhibited threatening behaviors without any hint of humor to dampen the inference. After a celebratory group gathering, he pointed out a bar on a street corner, telling me about a bar fight he had gotten into there a long time ago. In another instance, he recounted a dinner outing at a chain restaurant with his wife, explaining how he had intentionally parked his yellow Corvette, a possession he took great pride in and one that went right along with his machismo attitude, at the far end of the lot to avoid having other cars park near it. When they returned, the parking lot was nearly empty, but someone had decided to park next to his Corvette. This infuriated him to the point where he claimed to have taken a bat to the other car, just for parking beside him. I don't know if he thought these stories would somehow impress me, perhaps as a twisted form of courtship, but they were anything but impressive.

This was not the only way though that he instilled fear. Like he did just a month after I started, he would often lose his temper and lash out at people, yelling about work not being done to his standards or expressing his disapproval of what someone did or said. While it's acceptable, as a manager, to express disapproval and provide guidance to his subordinates, it should be done in a professional manner, not with digs, insults, and threats. Certainly not to the degree in which he used with me.

It was also hard to avoid this temper. He would change his mind in a way that could be construed as a managerial tactic but it was playing mind games and manipulating the situation while exuding microaggressions. For instance, if you were working on a task that had two options, if you chose Option A, he would tell you to switch to Option B, and vice versa. It wasn't about what was best for the job or client; it was about asserting dominance over you. Randall, whom I mentioned before as being good at deflecting with humor, would joke about it, saying, "You know Nick, it depends on which way the wind is blowing that day."

In addition to instilling fear, he also degraded me and undermined my confidence. He constantly belittled me by insulting my intelligence, another way of asserting dominance. While he could do this to anyone, it was much more prominent with me as a woman. It prevented me from feeling comfortable communicating around him and consequently hampered my career growth. These degrading comments were made both verbally and in writing, sometimes even in front of other employees in the company as well as clients or contractors, disparaging me and damaging my reputation and the overall success of the group. But he didn't just insult my intelligence. He diminished my entire being, treating me as if I were less valuable than the other group members and eroding my self-worth. It was cruel. As my time working with Nick continued, this fear, stress, anxiety, and depression grew, while my confidence, positivity, and overall well-being diminished. Nick was both verbally and mentally abusive.

DISCRIMINATION

I'm not sure what percentage of those in the workforce know their rights as employees, but I suspect that most don't. I myself was very naïve and unaware of my rights until towards the very end of my employment at HDR. Discrimination and harassment are serious issues, and we have civil rights which are meant to provide freedom against these issues as well as labor and employment laws to protect those rights. The Equal Employment Opportunity Commission (EEOC) is the federal agency responsible for enforcing these laws. Each state may also have its own laws and agency prohibiting additional types of employment misconduct.

The idea of "protected classes" mentioned above is important. If someone treats an individual differently or unfairly because of their membership in a protected class, it can be considered discrimination. These laws were first implemented in the 1960s during the Civil Rights Movement and have been evolving ever since. The initial purpose of these laws was to increase equality and protect people from the injustice suffered due to bias against their race, color, religion, or national origin. Several additional classes have been added since, including sex or gender. The current work-place rights movement begs the question of again adding to these classes or otherwise making adjustments for more equal rights and better protections for all, including against bullying and rights against neurotechnology.

Sticking to the current laws, the disparate treatment I endured from Nick was also a form of gender discrimination mainly due to his bias against women as well as against my national origin but I didn't realize the latter until after leaving the company. From the start, it was obvious that Nick treated me differently based on my gender, I just didn't understand my rights and the company completely took advantage of that fact. For example, Nick tried to delegate to me the office and managerial work that he didn't want to do, such as approving the group members' timesheets. That was his responsibility, not mine, and I was not comfortable with it.

When women began entering the workforce in greater numbers, many of them worked as secretaries. That was a long time ago, and it's discriminatory to assign a woman to do the secretarial work despite her other qualifications—or even to insinuate that she's somehow the automatic choice for such work. Among the members within the Tampa group, I was the most qualified and credentialed being a licensed Engineer as well as a qualified Team Leader on bridge inspection, which I get into more later. Although I was the only female in the Tampa movable bridge group, there was another woman who worked out of a different office and was also a mechanical engineer. She ended up submitting her resignation and Nick, finding amusement in the situation, said, "She left because they kept giving her spreadsheets to work on and treating her like a secretary," and then he laughed about it. It exhibited his ingrained bigotry and sexist bias.

Whether conscious or unconscious, it doesn't matter—it's discrimination, and it is not okay. Nick treated me as if I was incapable of performing my tasks, even though I was very capable and efficient. We used to have weekly meetings for the group. These meetings were meant to report on our progress from the previous week regarding work assignments and to estimate our projected hourly workload for the upcoming week. During these meetings, Nick would often treat me differently than the other group members. Instances like this caused the rest of the group to also lose respect for me. It was completely unacceptable, making my job more difficult.

Even just the tone in which he spoke to me compared to that which he used with the male employees was different—it was more aggressive. Often, if I made a comment or suggestion, he would immediately dismiss it, regardless of its validity. In fact, at times it seemed he wasn't even hearing me. Other times, if I made a mistake, he would blow it out of proportion while if one of the males made a mistake it typically wouldn't be treated as a big deal. It was clear disparate treatment.

THE IMPORTANCE OF DOCUMENTATION

Over the course of my employment with HDR, I continued to try and make things work. First, I tried using different approaches to the way I talked to Nick and how I responded to his negativity and degradations. Remember, Nick exhibited a "boss" managerial style, which meant he tended to micromanage, criticize, blame, and lead with fear. From my interactions with him, it appeared that some of this behavior stemmed from deep-seated insecurities. In attempts to make him feel more secure, I tried using phrases like, "Oh, that's a good idea, Nick." However, this strategy didn't seem to work and, if anything, it appeared to exacerbate his need for superiority and dominance over me. I also attempted to be very direct, though in the simplest ways possible, saying things like, "When you talk to me like that, it makes me feel bad." But it was when I ignored or rejected his sexual advances that his insecurities were most triggered and his reactions and anger most peaked.

Imagine a situation where he made a mistake. My pointing out his error would also incite his anger. It seemed as though he felt I was challenging his authority and manhood and he exuded an attitude of, "Who do you think you are?" However, I was simply a team member who noticed an issue that needed attention. Ideally, this should have been welcomed, but it wasn't—not coming from me. He would even go as far as to knowingly and intentionally refuse to correct his mistakes, allowing flawed work to be sent to the client without admitting fault or rectifying the issue. This was an abuse of power, manipulating circumstances and bending rules for his own benefit, to maintain control and to portray me as the one in the wrong. Being my supervisor, he had a direct impact on my job. My commitment to the quality of our work and products made this situation especially troubling.

Another trait that distinguishes a boss from a leader is the willingness to share knowledge. Bosses tend to hoard information, often out of insecurity that someone else might one day replace them. Nick, in particular, was reluctant to share his knowledge and preferences with me. There were times when I would ask him to explain why he preferred to use specific details in our designs, and his response would simply be, "Because I said so." This approach is unhelpful to everyone involved. It didn't aid in my skill development, nor did it contribute to increasing our efficiency for the benefit of the company and our clients. It became increasingly apparent, and I couldn't help but wonder if he had any intention of helping to advance my career at all. His reluctance was hard to ignore, and it was clear that he did not treat others this way.

Eventually, the stress began to take a toll on me. I started exploring other options and avenues for navigating through the misconduct. That's when I joined Toastmasters, a non-profit organization designed to help individuals improve their public speaking skills. I believed that this could potentially empower me to speak up to Nick and assert myself more effectively. Additionally, it was an opportunity to enhance my overall career and leadership skills. Later on, I joined another professional engineering organization and participated in several career growth and management training courses at

HDR. I was also encouraged to sign up for the HDR mentorship program. Some of these steps were my own initiative, while others were influenced by individuals within HDR.

The most powerful tool you have when confronted with a situation such as this is to document and preserve the facts. Confronted with sexual harassment, psychological harassment, discrimination, disparagement, and abuse of power, I was at a loss as to what more I could do at the time. Consequently, I tried my best to concentrate on the tasks at hand, telling myself, "Just focus on the work." Speaking up about sexual matters is particularly challenging. However, I found it easier to voice my concerns when faced with verbal and mental abuse from Nick, which although I didn't fully understand it at the time was psychological harassment and discrimination.

I remained silent for longer than I should have, driven by fear. But about a year and a half into my employment, I began to voice my concerns to others and started to document my experiences. Unsure of what the future held, I felt compelled to take notes on Nick's words and actions. It was already apparent that he was distorting the truth and manipulating situations, so I believed it was crucial to protect myself and my job. There is nothing wrong with documenting situations like these, but first learning and understanding your rights is extremely beneficial to protect yourself from them gaslighting, manipulation, and being taken advantage of. When documenting, it is important to note the five W's: Who, What, Where, When, and, most importantly, Why. From this experience, I learned valuable lessons about the importance of proper complaint procedures and documentation in holding employers accountable.

3

UPPER MANAGEMENT, INITIAL COMPLAINTS, AND MORE

MOVABLE BRIDGE GROUP HIERARCHY

HDR's movable bridge group extends to a national level. There are offices across the country housing movable bridge engineers. Additionally, there are structural, mechanical, and electrical movable bridge engineers located outside of Tampa with whom we would collaborate. The company advocates for the use of a work-sharing program, enabling coordination and collaboration beyond the confines of the local office—a practice that proves highly effective in the specialized field of movable bridges.

On my initial employment, I was hired by Laurence Delgado. For the first year or so, Laurence served as Nick's supervisor. In time, he rose to the position of national lead for the movable bridge group, paving the way for Mitchell Mathews to step in and fill the vacancy left by Laurence's promotion, becoming the regional movable bridge manager.

Laurence is a tall, slender man with a generally affable disposition. In my view of him as a manager, Laurence truly stood out as a leader. He treated everyone as individuals, fostering a team spirit within the group. His approach was both encouraging and respectful. As an electrical engineer, I frequently accompanied him on work trips to inspect movable bridges—a task typically undertaken by both a mechanical and an electrical engineer. One trait that stood out to me was his profound care for his family. He would take calls from his wife or daughter while we were on the road. Although I couldn't understand their conversations, as they were in Spanish, it was clear that he held a deep and sincere love for his family. His affection was transparent and heartwarming, making me feel genuinely happy for them. In contrast to Nick, Laurence was the last person one would need to worry about—or at least, that was what I initially believed.

Then there's Mitchell, or Mitch, as he is commonly called. A structural engineer based in the Fort Lauderdale office, Mitch struck me as a generally good-natured individual. He was always ready to lend a hand and often described himself as a leader. Mitch placed an emphasis on maintaining appearances, aiming to ensure that our group projected a positive image. However, his level of engagement with our team was somewhat inconsistent.

At the beginning, it took some time for him to get acquainted with our projects, as he was meant to take on a project manager role. Then, although he was hired for the movable bridge group, he took on a fixed bridge management role out of Fort Lauderdale. This set him back with regards to his involvement with the movable bridge work. He became extremely busy and overextended. I will also say though, he seemed well organized. We got along for the majority of the time we worked together and I assisted him on a few structural tasks. He once told me that he looks for people like me to have on his team because I would keep him safe. I was hard working, honest, and direct. Our relationship, though, changed by the end of my employment.

MY FIRST COMPLAINT TO UPPER MANAGEMENT

Around the fall of 2020, I was assigned a role as construction inspector on a bridge contract near Richmond, Virginia for an extended period. The project encompassed both mechanical and structural work. Up until that point in my career, I had never worked as a construction inspector—a role typically reserved for specialized groups. In the process of my career, I had heard a few stories about engineers clashing with contractors on construction inspection jobs, sometimes resulting in the engineers being kicked off the job. Understandably, I had slight hesitation, but more so excitement about the new challenge ahead.

Despite my lack of experience in construction inspection, Nick provided me with little to no information or guidance on what to expect or how to approach the task—despite the fact that I had explicitly informed him of my inexperience in this area. Left to my own devices, I searched the company server for information about the project, finding helpful materials referenced by another mechanical engineer from HDR's Virginia office.

And so, I made my way to Virginia, ready to tackle the vertical lift style bridge assigned to me. This bridge design features a main span flanked by two large towers, with the span moving vertically along the edges of the towers through numerous sets of ropes and sheaves—components similar to pulleys, powered by machinery systems. The ropes are fastened at one end to the span, and at the other end to a counterweight within each tower, allowing for an efficient system and balanced movement: as the span raises, the counterweights lower, and vice versa. The ropes, subject to replacement after a number of years depending on their type and the load they carry, are closely monitored to ensure they are replaced prior to or nearing the end of their lifecycle.

Replacing the ropes was part of the mechanical work to be performed under this contract. This is a significant task, requiring jacking of the span. Along with the replacement of the ropes, the additional mechanical work primarily involved replacing some existing clutch couplings in the main drive machinery systems, which were necessary for rotating the sheaves and operating the bridge. On the structural side, outriggers or steel supports were installed at the top of the towers around each corner of the main vertical lift span. These supports facilitated the hanging of new catenary cables across and overhead the span.

Initially, I experienced no issues while working on this assignment. My interactions with the contractor were positive; they were generally respectful, responsive, and cooperative—contrary to any adverse stories I had heard. A crucial aspect of construction inspection is the instruction to avoid "directing work." This request is in place to prevent liability issues; if problems arise with the performed work, it prevents a blame game. In my opinion, the most challenging part of this task is restraining yourself from intervening when you foresee potential issues. It's like watching a horror movie with your hands over your eyes, wanting to look although you know you shouldn't. As a construction inspector, you are basically expected to allow the contractor to make mistakes, even if they unfold right before your eyes, costing time and money—unless, of course, it's a matter of safety.

Despite this, I continued to work well with the contractor, staying on the project for just over a month without issues. That was until Nick's arrival. During a weekend closure for the clutch coupling work, which necessitated additional inspectors, Nick, myself, and the male mechanical engineer from the Virginia office were all present, alongside the contractors. Throughout this period, Nick's behavior towards me was starkly different from his interactions with the male engineer. He yelled at, belittled, and ridiculed me, both in private and in front of the contractors, whereas he remained respectful towards the male engineer. His disparate treatment and bias were very clear. I could also sense jealousy from Nick of my working rapport with the contractor, stemming from his inappropriate and sexual interest in me. He manipulated the situation to his advantage, sabotaging me and ensuring that the contractors lost respect for me—and it definitely worked.

This treatment completely changed the dynamic of the relationship that I had built with the contractors. They knew Nick was my boss; if he treated me that way, why couldn't they? This was pure disparagement. For the remainder of the weekend, and after Nick's departure, I no longer commanded respect from the contractors, which significantly hindered my ability to perform my job. They became unresponsive and went as far as to taunt me, placing cones in front of or on top of my car. They promised to pick me up in their small boat to take me to the barge they were working from, only to leave me stranded at the dock. They misled me about their location on the bridge, claiming to be in one place while actually being in another. These issues hadn't arisen before Nick's arrival.

One day, I observed one of the contractors deviating from the specifications laid out in the design plans. I promptly informed the contractor's project manager, following the proper chain of command. He concurred with my assessment. However, after a considerable wait, the issue remained unresolved. Once again, I found myself being ignored, a result of Nick's previous mistreatment. The situation was incredibly unfair, but I was determined to do the right thing. When my subsequent attempts to reach the manager proved unsuccessful, I approached one of their workers for assistance, which ultimately was used as an excuse resulting in my being kicked off the job.

The night following Nick's disrespectful and disparaging behavior on the bridge, I returned to my hotel room, overwhelmed and distraught, desperate for a way to address Nick's nasty treatment. I decided to call Laurence looking for support in handling the situation and to disclose everything that had been transpiring. Throughout our conversation, I was crying and obviously upset. Laurence inquired if I considered Nick's behavior to be harassment. Unaware and naïve about my rights, not yet having sought legal counsel, I exuded doubt. The mere mention of harassment made me anxious, fearing it might exacerbate the situation. Laurence quickly changed his tune, describing Nick's actions as merely "harsh"—his choice of words. My being naïve also made me very gullible to his now obvious downplaying on the severity of the situation and prejudicial objective to protect Nick, when I was the one in need of protection. This was my first complaint to upper management and the beginning of what would be an appallingly long trail of gaslighting.

WORK WITHOUT PAY AND DISPARAGED AGAIN

Nick often belittled me, as I have already mentioned, through degrading comments and insults. However, it was particularly disparaging when he did this in front of others, especially in front of or directly to clients. Sometimes his behavior stemmed from aggression and sexist tendencies, while other times it involved manipulating the situation and abusing his power. One such instance occurred in Florida during a routine inspection contract covering multiple bridges. HDR had been contracted by another engineering firm to provide mechanical and electrical expertise for the movable bridge inspections. This collaborative approach is common in the bridge and movable bridge industry, where two engineering consulting firms join forces on a single contract. Typically, one company serves as the prime consultant and the other as the subconsultant. In this case, we were the subconsultant. This particular contract though was set up with very limited hours and as such Nick had asked us to work without pay if we didn't have enough hours. Yet another instance of poor management and illegal actions.

For an extended period, the prime consultant had conducted these inspections independently, employing another inspector. However, when that inspector left, HDR stepped in to fill the void. Unfortunately, the former inspector did a poor job on the previous inspections, resulting in rather sparse and inadequate reports. As a consequence, many of these bridges had not been properly maintained for years. When I took over the responsibility, I worked to inspect the bridges up to standards and reported actual representative conditions of the bridge components. Not only were the conditions now accurately depicted, but the equipment would also be better maintained. Alongside reporting conditions, we provide maintenance recommendations. After all, the primary purpose of performing an inspection is to create a historical representation of the bridge and maintain good condition of its components.

One day, a few months after I began working on these inspections, I overheard Nick taking a call near the office windows—a frequent occurrence. He was speaking with the prime consultant's project manager, who seemed to be questioning my level of detail in the inspection reports, as it differed from what he was accustomed to. Nick's response was to disparage me, telling the project manager, "She doesn't know what she's talking about." This statement was not only insulting but also served to undermine my credibility. Nick was both negatively impacting my work and my professional reputation.

Based on the accounts of others who had already worked with this project manager, as well as my own experiences, he could be excessively demanding. At times, he would follow me around on the bridge, trying to rush me through the inspections. Again, further loss of respect, being influenced by Nick's comments. It was yet another example of how the derogatory statement about me made my job more difficult than necessary, and creating unsafe working conditions on the bridge for the inspection team. I eventually brought this up to Nick, given that he was my boss and that employers are obliged to ensure safe working conditions for their employees. Right?

Nick advised me to inform the project manager that HDR was contracted to provide specialized engineering services and that we would deliver those services according to our standards. While it initially seemed supportive, it turned out to be a trap. He also mentioned that the hours allocated for this job were insufficient, explicitly stating that anyone needing more time would have to work unpaid.

During the next inspection, I encountered the project manager and conveyed Nick's message about adhering to HDR's standards. This upset him, leading to a conversation between him and Nick, after which I was removed from the job.

My intentions were genuine, driven by my concern for the bridge's reliability and safety. However, Nick threw me under the bus. His true intentions became increasingly apparent: he aimed to push me out of the company. A few weeks later, another member of the prime consulting firm inquired about the upcoming inspection and who would handle the mechanical aspects. Nick suggested Thomas as my replacement. Initially, the prime consultant seemed hesitant; however, Nick eventually persuaded them, resulting in Thomas taking over the role. Both Thomas and Marcus at points mentioned to me that they had been contributing a significant amount of their personal time to this contract—a practice that is unequivocally unacceptable. Employees should never be compelled to work off the clock. To mitigate the unpaid labor requested of me, I simply tried to work harder and faster.

HE THREATENS MY JOB

After Nick's disparaging comment to the prime consultant in Florida, and right around the time of his visit to Virginia, he also threatened my job via email. This incident occurred on a rehabilitation project that Mitch had taken over as project manager. For any new or rehabilitation designs, there are typically multiple submissions sent to the client, such as a 30%, 60%, 90%, and 100% submission, followed by a final submission. This process provides the client numerous opportunities to review the work or to have it peer-reviewed by another consulting firm. Once their review is complete, we receive their comments and are then given a timeframe to respond to each one.

Mitch, having taken over the project, was proficient at keeping things organized. He had emailed out the 100% submission comments we received back from the client so we could provide our responses. I had discussed this with Nick, and we agreed that I would send him the responses for review before sending them out. However, I had also spoken with Mitch who wanted an update on our progress since two weeks had already passed since he sent the email. Consequently, I drafted the responses, emailed them to Nick, and CC'ed Mitch. This was standard protocol. I made it clear in the email that the responses were for Nick's review. Almost immediately, I received an email back from Nick threatening my job. He stated if I continued to ignore him, which I wasn't, "you will find yourself not working for me any longer. Understood?" It was an entirely unreasonable response... and it was harassing.

This episode represented "severe" harassment, leaving me extremely upset. I was terrified that I would lose my job, and worried he might actually fire me simply for CC'ing the project manager on an email—a standard practice. His reaction was utterly unacceptable and absolutely unnecessary. Distraught at that time, I called Nick after receiving the emails. I pleaded that it was unreasonable to respond to me in such a manner, expressing that his job threats were not okay and that he frequently treated me in an aggressive and degrading manner. He focused exclusively on his desire to review documents before they were sent out, dismissing my concerns. I explained that I had explicitly stated in the email that the comment responses were for his review, not for Mitch to distribute. One of the many instances he didn't take in what I was saying or writing to him, rejecting to see or believe in my capabilities. Nevertheless, he failed to acknowledge the error in how he had responded.

This incident is what made me start documenting any intimidating, offensive, threatening, unfair, and unreasonable things Nick did or said to me. His mistreatment forced me to protect myself. I assumed if things got worse at least I would have a list of what he was doing to me. Better yet would have been if I had already understood my rights.

As I noted previously, if someone is harassing you or discriminating against you, documenting the events is extremely important as well as retaining a copy in writing. Keep a notebook or journal of each incident, even if no one but you and the harasser are involved and remembering the 5 W's: who, what, where, when, and why. I included things like dates, names of people involved, details of what was said or done to me, and what job it was for or where it happened. You may also find yourself including the feelings that these incidents provoked in you, which is also okay as it is important to document how the conduct is affecting your mental and physical well-being. Is it causing you stress? Is it changing your mood or behavior? Is your work suffering as a result or are you less productive? The 5 W's are the facts and what you will be asked for later on, but how it affects you is also important. For me, I actually started working harder and became even more productive from the stress that Nick caused me. I was in fear of losing my job so I figured if I did more it would protect me from getting unjustly fired.

MORE ABUSE OF POWER AND QUALITY CONTROL

The transition to a management role often means doing less of the technical work. Instead, your role shifts towards assisting others in performing their tasks. This is a key aspect of creating succession within a company. If executed properly, it can significantly enhance employee motivation, encouragement, and retention, leading them to stay with their current employer. Boosting employee retention in this manner not only saves the company a considerable amount of money, but more crucially, it fosters a positive and safe working environment that permeates the entire team. However, if a manager resists this progression, it can detrimentally affect the employee, the team, and the entire organization.

This was part of the issue with Nick. He was reluctant to share his knowledge and experience, particularly when it came to me. While some individuals might be content with this situation, merely working for a paycheck without giving much thought to the future, I did care and wanted to continue to grow. I am career-oriented, always keen to take on new tasks, and looking to continually expand my skill set, and eventually wanted to advance within the company. Consequently, working with Nick required me to exercise a great deal of patience.

We once worked on a rehabilitation project for a bascule-style bridge, which operated using a hydraulic system with cylinders serving as the prime movers to open and close the bridge. The project mainly involved replacing the Hydraulic Power Unit (HPU) and the cylinders. Nick took on the responsibility of performing the design calculations for the hydraulic system. He did, however, ask me to review his calculations. As I have mentioned before, he did not appreciate it when I pointed out mistakes in his work. Sometimes, Nick would make flat out arrogant remarks, such as "I did it, so there are no mistakes," which is an absurd and narcissistic attitude. Everyone makes mistakes.

In our industry, we implement a quality control (QC) process. Essentially, this process ensures that due diligence is exercised to correct errors. First, the person who performs the work hands it off to a reviewer. The reviewer then provides comments on the work. These comments are subsequently reviewed, either accepted and incorporated, or rejected if deemed invalid by the original worker. Finally, an updated version of the work is checked against the review document to verify that all comments have been addressed. As engineers, this process is important.

Since I was the reviewer in this instance, I reviewed Nick's work and sent my comments to him. He accepted most of my comments and rejected one of them. In the final stage of the process as verifier, I emailed him, inquiring about his rationale for rejecting it. The comment in question addressed a hydraulic value he had used in the calculations, which he had pulled from a manufacturer's data chart. Upon reviewing the calculations, I noticed the chart had been misread, resulting in an incorrect value in his calculations. This is an easy mistake to make, and it underscores the importance of the QC process. Despite this, he refused to acknowledge it as an error. Attempting to remain polite, I asked him to "explain for my understanding." My intention was not only to maintain civility but also to avoid provoking any hostility from him.

Nonetheless, he attempted to shift the blame onto me, gaslighting and making it seem as though I was the one mis-reading the chart. This not only undermined my confidence by introducing doubt but also created a false narrative. Such behavior is detrimental, not just for the current job, but also in the long term. It is the antithesis of fostering succession, a tactic he purposefully employed many times through the distribution of misleading information. Eventually, he did acknowledge his mistake in an email, yet he still refused to amend the calculations. His response was to tell me to "DROP IT," and so, reluctantly, I did.

Just about a week later, we were reviewing the plans for the same project. Both Randall and I had been working on the plans, and Nick was the reviewer. In addition to the first review, we referred to as the "detailed check," there is a sec-ond quality review that looks at the project as a whole. In this case, Nick had already performed the detailed check and then tasked me with performing the second review. However, when assigning me the task, he specifically told me not to spend a lot of time on it because he had already checked it—once again, unwilling to admit that he could make a mistake. I rushed through the review as quickly as I could and still found a few minor things. I sent Randall the comments with Nick CC'd on the email. Nick responded via email, rejecting one of my comments, and stated that it was what he "specifically gave Randall to write down," followed by "Quit changing things you don't understand." It was insulting and degrading.

The review comment that I made, which he was rejecting, was on our hydraulic component tables. These tables are often listed in plans to include identifications of each hydraulic component, the quantities required for each component in the contract, and additional descriptions and ratings for each component. The tables are used in conjunction with and correspond to hydraulic schematics also shown in the plans. This relays the overall intended hydraulic design from the HPU to the cylinders with all the different valves, filters, hoses, piping, etc., in between.

While performing my review, I noticed that there were two valves which were part of the same subsystem; however, their descriptions in the tables did not correlate well. That was the intent of my comment. It may have been minor, but that doesn't mean it shouldn't have been addressed. When we create contract documents such as plans, they are kept for the life of the bridge and used for reference in the future. Not addressing errors or comments made for clarity is a disservice to the client. After speaking with Nick on the phone, he finally understood why I made the comment, and this time the plans were corrected. But this was only after he had already insulted me via email with Randall CC'd.

These types of instances with Nick were disheartening. Not only that, but it was degrading to me when he'd put me down in such a way, hurting my confidence. The accumulation of this type of treatment from Nick was "pervasive" harassment. By this point I had already called Nick out directly and complained to Laurence about him. Soon I would have another opportunity to talk to Laurence.

MY SECOND COMPLAINT TO LAURENCE

About a month after I first complained to Laurence from the hotel room in Virginia, I was back on a routine inspection in Florida with him. I again attempted to talk to Laurence about Nick and his behavior. We sat in the car next to the bridge, talking for a while, as I was once again upset and even tearful explaining it to him. He kept making excuses for Nick's behavior, saying, "That's just Nick." He was ignoring the impact Nick's behavior was having on me and how it was affecting me, though he could plainly see it right in front of his eyes.

I now recognize it as both harassment and discrimination but, at the time, I was unaware of the law and the importance of using words like these. The current laws require the behavior to be about a protected class for you to be protected by law; even when it is not explicitly or overtly revealed. As managers, they know this but as an employee I was completely uninformed. I didn't know what I had to say or put in writing to protect myself. It was because I am a woman although I mainly focused on his actions. I always knew it, but I kept focusing on what he was doing and how it was not okay. It's as if we are programed not to say the words or phrases that can protect us the most. With the conversation not really going anywhere, we decided to start the inspection. I wiped away my tears and got to work.

Since Laurence was avoiding addressing the real problem, I started to realize that I would have to talk to someone else about Nick. What they did to me was no doubt intentional and negligent. Looking back, I can't help but think about how close and loving Laurence was with his wife and daughter. How would he feel and what would he do if someone was treating one of them this way? Would he think it was ok? Of course not.

I START TALKING TO MITCH

After complaining to Laurence twice and not getting anywhere, I decided to try talking to Mitch. He was Nick's new supervisor anyway. Seemingly helpful, Mitch was more willing to spend time discussing the issues with Nick. Mitch would say that he and I got along fine, followed by "it's not like I have to give you flowers or anything." This was a figurative way of saying he didn't need to compliment me or do things of that nature for us to get along. He would make comments like "I can do this" or "I can mediate." He seemed sincere in his desire to help Nick and I resolve our issues, so I was hopeful.

When I would complain about Nick's behavior, I often used the word "aggressive." When I said this to Mitch, he responded by calling Nick "gruff." This was similar to Laurence saying "harsh" in my first complaint. They constantly downplayed my complaints and sometimes just glossed over or ignored what I was saying. Some other words I would repeatedly use were bias, angry, and fear. I was scared of Nick, and yet they just let it continue.

While you are right in the middle of it, sometimes it is harder to see what is happening over the course of a long period.

Incremental effects of day-to-day interactions are much harder to see than when you compare where you started to where you finish. That is why documenting everything is so important—it's like a journal you can look back at. Everything that Nick was doing, little by little, was hurting me; it was abuse. I tried so many times to ask for help, but when the people you reach out to take the harasser's side, they can work to manipulate and gaslight you, or twist your complaints against you instead.

Now and then I would reach out to Mitch again about Nick, typically after a more severe incident or after the accumulation of incidents got to me. I know that Mitch would also have conversations with Nick about this, but I don't know the details of those discussions. It's possible Mitch was making Nick aware of how his unfair treatment was affecting me and that he needed to stop, but it's also possible their conversations revolved around bad faith.

But the harassment and discrimination did not stop, and things were in fact getting worse. It was not okay for him to communicate and mistreat me in this manner. The repeated occurrences were eroding my confidence. I believe it is this type of degrading treatment that causes what some might call imposter syndrome, suing doubt into someone with repeated putdowns and insults. By this point, I had been working under him for almost two years, and the cumulative effect was definitely catching up to me.

"NO ONE CARES ABOUT YOU"

Another notable aspect of Mitch was his generosity. If you asked him for something, he would go out of his way to make it happen. During my particularly challenging times dealing with Nick, and in my discussions with Mitch about it, we explored other options. One such opportunity Mitch provided was a chance to serve as the project engineer on a newly acquired contract. This contract entailed a major rehabilitation job on a movable bridge in Fort Lauderdale, encompassing significant work across all three disciplines: mechanical, electrical, and structural. I had not yet taken on the role of a project engineer, so I was eager for the new challenge. A project engineer is responsible for keeping the team on track and ensuring that work progresses towards the ultimate goal of delivering on the designs for the agreed-upon scope of work items. This role collaborates closely with the project manager, whose primary responsibilities include client relations and managing the project's financial budget. Since joining our team, Mitch typically filled the project manager role, as he did for this project.

This role also provided a slight reprieve from working with Nick, which was a relief. My background in both mechanical and structural engineering coupled with my overlapping work in the electrical discipline gave me a significant advantage in this role. Many professionals in our industry specialize in one of the three disciplines and continue in that single path throughout their careers. While there is nothing wrong with that approach, and many find great success in it, I was interested in understanding the system as a whole allowing me to transcend the boundaries of the three disciplines as needed.

From the company's perspective, this flexibility is highly beneficial. For instance, if the mechanical department had a light workload, I could be assigned a structural task or asked to review something for the electrical team. This adaptability made me well-rounded in the industry and particularly well-suited for the project engineer role, which encompasses all the disciplines.

At the outset of a rehabilitation project, it is immensely beneficial for those involved in the design to conduct a field visit and inspect the bridge in person. This practice offers a clearer understanding of the bridge's condition leading to increased efficiency in subsequent office design work, ultimately saving time and money. For this major project, which engaged all three disciplines, we conducted an inspection with several team members present. In attendance were Nick, Laurence, Mitch, Marcus, myself, and an additional colleague from the Fort Lauderdale office whom Mitch had invited.

Upon our arrival at the bridge, we got off to a great start. The weather was perfect, typical for Florida in the spring. The bridge itself was a bascule style featuring existing electro-mechanical main drive machinery, though it lacked redundancy in its system. Many movable bridges, similar to this one, have been upgraded to include redundancy aiming for uninterrupted operations in the event of a main motor failure. Without such redundancy, a bridge may need to be operated by auxiliary systems, backup power, or even manually cranked, potentially resulting in significant operational delays and traffic congestion.

Our primary goal during the site visit was to assess the conditions of the existing components, verify critical dimensions, and brainstorm potential recommendations for the rehabilitation designs. While we anticipated replacing some elements, others depended on the client's budget. One component slated for replacement was the rotary cam limit switches, commonly used in movable bridges within the controls systems. These switches track the bridge's position during operation by coupling to a shaft within the main drive. As the shaft rotates, the limit switch sends a signal to the control system, converting the number of shaft rotations to represent the bridge's position in degrees as it rotates open in the case of bascule bridges.

The rotary cam limit switches are a point of overlap between electrical and mechanical engineers. Electrical engineers are responsible for integrating them into the control systems, while mechanical engineers choose the appropriate size of the limit switch based on the shaft to which it is coupled. Additionally, mechanical engineers design the supports and anchors for these limit switches. So, as we were in the field, Nick and I were discussing the locations of the existing limit switches and determining where to install the new ones. After making a decision, we went our separate ways.

About 30 minutes later, I rejoined the group and found them discussing their findings, focusing particularly on the placement of the new limit switches. I chimed in, sharing the decision Nick and I had reached earlier. To my surprise, Nick responded, announcing, "No one cares about you." It seemed he and Laurence had discussed the matter further and decided on a different plan. While changing plans is perfectly acceptable, there's no need to personally attack others in the process.

HEALTH EFFECTS

The comment haunted me, occasionally resurfacing in my mind: "No one cares about you." Such a hurtful statement. It was fortunate that this incident occurred at the end of the inspection, as I needed to retreat to my hotel room. After already suffering years of degradation from him, the remark took a significant toll on my mental wellbeing.

But, not only was Nick degrading, he rarely ever gave positive feedback. It can be shocking how much one's confidence can be boosted with simple affirmations like "You got this." Phrases which can be delivered without much effort although surprisingly uncommon in many people's vocabulary.

A few days after the comment, I complained to Mitch about it. When it happened it was in front of the whole group that was out in the field. Every single one of them put their heads down after Nick's statement. They all knew it was wrong. Even Nick knew it, as you could tell from his face as soon as the words left his mouth. But again, Mitch downplayed it as though it wasn't a big deal. It was a very big deal to me.

As mentioned previously, I already noticed negative effects around this time to my emotional, mental, and physical well-being. I felt a significant loss in concentration. Additionally, I became depressed and started to feel extremely lonely. I started eating more. I was raised in a household where we were taught to eat healthy. This was completely different and I gained noticeable weight over the course of a few months. At this point, I again knew it was time for the next step, and that I would have to reach out over Mitch's head and out of the movable bridge group.

4

HR IS NOT YOUR FRIEND

"HUMAN RESOURCES" OR "COMPANY RESOURCES"

If you work for an employer, you should have access to a human resources (HR) department. But what exactly is the purpose of this HR department? Judging by its name, you might think its role is to assist the employees. After all, you are human, so why wouldn't a department with "human" in its title be there for your benefit? This, however, is a common misconception that many companies inadvertently perpetuate and take advantage of.

HR primarily exists to serve the company's interests, not the individual employee's. They oversee the hiring and firing processes, and their intermediary activities largely revolve around protecting the company's wellbeing. So, the next time you consider approaching your HR department, bear in mind: HR is NOT your friend.

This isn't to say that you should never communicate with them. They can provide valuable information and answers regarding company-mandated benefits, for example. They should also be a reliable source for inquiries specific to company policies. Whether they consistently uphold these policies is, unfortunately, another matter entirely. Depending on your company's structure, there may be separate departments designated for specific issues, relegating HR to a directional role. Essentially, they should be viewed not as a "Human Resources" department, but rather as a "Company Resources" department.

It's akin to visiting a website and clicking on the frequently asked questions section or calling a help line. But if that's the case, one might wonder why we need human representatives in company HR departments. While they might assist in employee training or perform risk management duties, this involvement with "human" aspects can be misleading. The portrayed intension for this department is to help in resolving personal matters between employees in work-related issues. While it would be great if this were their primary objective, the false impression can unfortunately leave naïve employees subjected to further abuse. Although HR representatives may display concern and compassion on the surface, their primary focus is on assessing potential risks and implications for the company, both in the short and long term. They are not your friend, ally, therapist, or confidante. Sharing intimate details with them, especially when emotional or upset, is generally inadvisable. They may appear supportive in the moment, but after you leave the room, their attention shifts from your wellbeing to risk management and the potential impact on the company.

What you could do, for the next time you decide to talk to HR, is plan ahead. If you have questions, make a list of what you want to ask and stick to that list. Even better, you could send it in an email. But if you must have a conversation, be careful when answering prying questions that they interject, because they are assessing risk. They will use whatever info they can against you if it comes down to it. They should not be trusted blindly. In my opinion it is very likely that HR will eventually be replaced with Artificial Intelligence (AI), either partially or fully. They have already integrated AI into systems. Regardless, you could start thinking of them now in the same way as AI: as a resource—in this case, for company information and a place to report specific grievances while knowing that it will analyze your reporting's possibly even still with bias or misunderstanding.

INITIAL CONTACT WITH HR

Within the few weeks following the Fort Lauderdale field visit, during which Nick made the terrible "no one cares about you" comment, I decided to reach out to HR. Our HR representative in Tampa was a woman named Jessica Howell. Like most HR representatives, she came off as polite, soft-spoken, and inviting. She seemed like a kind and considerate person with good intentions. However, I ultimately realized that she was far from it. Part of her job required her to present a façade. She manipulated and made me believe she was supporting me, and while she did offer some advice, her main priority was protecting the company.

We scheduled a meeting, and I prepared a list of points I wanted to discuss. Overall, I talked about how I had been struggling and wanted to open up about my experiences. I asked if it was possible to speak confidentially without filing a formal complaint at that time. I didn't want things to get worse. I also didn't want him to lose his job or something; I just wanted the situation to improve. Jessica, while being proficiently trained in these matters and fully aware of the company policies, outwardly simply responded with something like, "Of course." I explained that Nick was my supervisor, and had been for almost two years. I conveyed how difficult it was to work with him due to his often harsh demeanor. I emphasized that I was content with HDR as a company, but Nick's discouraging behavior was causing me to worry about my future there.

So, where did I go wrong? First of all, I didn't have to mention anything about whether it was a formal complaint or not. I could have simply provided the information and allowed Jessica to handle it as required. However, I was afraid of getting Nick into trouble and potentially provoking even more aggression from him. My second mistake was my choice of words—I used the term "harsh." Did you notice? When talking with Laurence and Mitch, I would describe Nick as aggressive. But they downplayed it so many times that I eventually began to echo their terminology. He was much more than just harsh.

But then she posed an investigative question, seeking additional information. She asked me to elaborate on what I meant by "harsh." I told her that it was hard to trust him. I explained that he would put me down, failed to give me credit when due, and threatened my job over minor matters. I also mentioned that when I asked him to justify his approach or explain his rationale, he would dismissively reply with, "Because I said so." Interestingly, she chose to gloss over the other points and zeroed in on the last one, asserting, "It's his job to answer questions." However, the belittlement and threats were, in fact, the most crucial aspects of my statements. In hindsight, I am certain those were the elements that should have been most concerning.

Ultimately, I conveyed enough information to at least consider warranting protection from the company, and under the law but they neglected to do so. Harassment is defined by acts of intimidation, degradation, and hindering professional growth. The only additional criterion required by law is that such treatment is discriminatory, based on one's belonging to a protected class. I needed to assert that this was happening because I am a woman. Indeed, it was the case, but at that time, I was unfamiliar with the legality of it all. Nevertheless, if she were genuinely acting in my best interest, she should have inquired about this aspect. Instead, her actions were aligned with safeguarding the company's interests and avoided the topic.

If, or when, all of this transitions to an AI system, it will be imperative for the system to pose all relevant questions— how could it not? Furthermore, the entire exchange would be documented. This brings me to my most significant oversight: I failed to send her anything in writing post-meeting. In retrospect, I should have composed an email summarizing our discussion and sent it to her. Putting it in writing would have formalized my complaint, complete with my signature, but at least I still had my notes.

CONDUCT CONTINUES

Although I had spoken with Jessica from HR, I remained uncertain about whether she had communicated with Nick regarding the issue. I was unsure if my message had been relayed to him, or if she had taken my complaints seriously. Regardless, the mean-spirited, aggressive, and threatening behavior persisted.

A couple of weeks after my conversation with Jessica, Nick summoned me to his desk. We were discussing a task he had assigned to me as he reviewed some design plans I had created. His tone quickly became disrespectful. Disturbed, I stood up from my seat, only to have him yell, "Sit the f@*k down." This outburst occurred in the office, within earshot of our colleagues. One nearby coworker even stood up, alarmed, to see what was happening. Reflecting on it now, I realize that around this time, I gradually began standing up for myself, as evidenced by this particular incident. I leaned down, addressing him at eye level since he remained seated, and responded firmly, "Don't talk to me like that." I then stood my ground, arms crossed.

However, the derogatory behavior resumed a few weeks later. Mocking me during an instant messaging exchange, he referred to me as "Bogdan," a former coworker he frequently ridiculed for his soft-spoken Russian accent among other things. This was both a putdown and an act of disrespect, entirely unnecessary. This is one of the instances that showed he also had a bias against those with Eastern

European national origins as he targeted individuals from Russia, Poland, and myself from North Macedonia. I didn't realize this until much later having taken a seminar on cultural differences. Nonetheless, and despite these ongoing incidents, I began to document as much as I could.

HOSTILE WORK ENVIRONMENT

As the stress of the situation between Nick and I rose, the abusive and uncomfortable working conditions continued to escalate. Was this a hostile work environment? I continued to voice my concerns about the ongoing issues to both Mitch and Jessica, but they consistently downplayed the situation, showing little concern for my wellbeing. Hostile work environments not only place stress on those directly affected by the negative conduct but also creates tension for bystanders who witness it. It fosters an unsafe and unfavorable work environment, from which we should be safeguarded by law.

After two years at HDR, I found myself overwhelmed with stress. I found myself more reactive with accumulated emotions generated by frequent comments like, "You don't know what you're talking about" and "No one cares about you", as well as from feeling demeaned, intimidated, and threatened. Do not let anyone victim blame you for your reactions to their abuse. Despite my repeated complaints about his behavior, it seemed like no one was genuinely intervening to protect me from Nick.

The indifference of HR and upper management towards these complaints also makes them culpable in enabling these conditions to perpetuate, leaving them further open to liability with negligence. Their lack of ethics and accountability in addressing these injustices fosters a climate of mistrust, feeling unheard, and a lack of belonging within the company, all of which were carried out in bad faith.

INTERJECTING POLITICALLY CORRECT LANGUAGE AND THE GROUP PERFORMANCE EVALUATION

The stress was twofold. On one hand, I was grappling with Nick's discriminatory and harassing misconduct towards me, and on the other, I was worried about my own reactions and uncertain about what might happen next. Two years after the start of my employment, there was a noticeable change in me. I prefer and value kindness and respect. Conflict is something I would rather avoid when possible.

Eventually, we had our first meeting aimed at addressing the issue, or at least that's what I believed at the time. Organized by Mitch, the virtual meeting also included both Nick and I. During the meeting, I was oblivious of their intentions but in due course it became clear that they had come prepared. They began to interject politically correct language and comments designed to protect the company and minimize liability. They made statements like "Nick treats everyone the same," an indication that they were aware of and covering up the discrimination at play, since he clearly did not treat me the same.

Discrimination or harassment can happen facially, where the misconduct is overtly against someone based on a protected class, committed with explicit derogatory and offensive statement or slurs based on gender, race, religion, or age for example. However, the misconduct can also be non-facial, more covert, and inexplicitly seem neutral on its face. This becomes harder to prove, especially when an employer works to cover it up. Because I did not understand my rights, I was mainly bringing up Nick's inexplicitly harassing actions, although there was also more overt offensive statements and actions he was making towards women in general, as well as the direct sexual harassment. What they were doing in this meeting was trying to redirect the behavior as far away from both facial and inexplicit discrimination or harassment as possible by working to gaslight me, as if Nick wasn't treating me differently. It was done intentionally, even if it meant they were making Nick out to be a "bully", just not a "harasser".

At this point, Randall had already left the company after buying some real estate and this was another interjection they made. They tried to make it as though Randall had left the company because of Nick running him off. Nick, however, through his engrained sexism, proceeded to say that he "pushed Randall to motherhood" and that now he was a stay-at-home dad. This meeting encompasses the full problem we were faced with. I didn't understand my rights, Nick couldn't see his own bigotry, and the company was working to cover it up. Even in the mists of the meeting he was being sexist.

They also attempted to lead me into making statements by posturing questions that would shift the blame away from

Nick and onto me. Mitch, for instance, asked me if I agreed that the communication issues with Nick followed a certain pattern, framing it in a way that suggested I was simply unwilling to listen to Nick. I took a moment to pause. The phrasing of the question made me feel uneasy and somewhat suspect of a setup, though I was not entirely sure of their motives. Thankfully, I did not take the bait and instead expressed that I felt Nick was not listening to me and tended to dismiss what I had to say.

During the call, I found Nick's unusually polite and amicable behavior almost surreal, especially given the stark contrast to his usual demeanor. They must have prepared him for this meeting. Meanwhile, I was simmering with pent-up frustration, exacerbated by Nick's fake act, which only served to heighten my aggravation. I even pointed out during the meeting that his current behavior was not reflective of his typical demeanor towards me.

In the meeting, we also first discussed them implementing a new system to track information on our group and our work. Mitch had been in communication with an administrator to gather data on charged work hours for various projects, all pertaining to the members of our group. Throughout this discussion, Nick made several comments insinuating to his lack of faith in my abilities, with remarks such as "I can't do the work for you." This marked the beginning of the creation of a Group Performance Evaluation for our group's members. They began to monitor and track the assignments given to us, the time each task took us to complete, and the quality of our final work. Why was all this necessary? It stemmed from Nick's sexist bias against me, rooted in his belief that I was a costly burden to the team and the company. Ultimately, this Group Performance Evaluation served to prove just how valuable I truly was, evidence of the discriminatory nature of the situation.

At the meeting's conclusion, Mitch stated his intention to schedule a follow-up meeting a few weeks later. He also expressed the need to reach out to HR, admitting that the situation had exceeded his expertise and that he required additional information or guidance. However, I now wonder: was the intention to seek advice on mediating the problems and resolving them, or was he being coached on how to steer the discussions in a way that protected the company at the expense of my protection from harassment and discrimination? Their actions make it clear that they were not operating ethically or within the law.

FOLLOW UP WITH JESSICA

A few days after this, I had a call with Jessica. It was partly because I felt confused by the last meeting due to the way they acted and the things that were said, as gaslighting will do. I was also looking for additional guidance from her. While on the call, she seemed to almost make light of the situation and was somewhat trying to laugh it off, downplaying and dismissing the severity of it. No part of this was funny to me.

Jessica also began to gauge my understanding of employment law and discrimination rights. My being completely unaware, it went right over my head at the time. She brought up a story about a prior situation she dealt with that had to do with racial discrimination. The story had to do with a supervisor calling a spare workstation a "slave station" and another employee being offended by it. As ridiculous as it seems to me now, I could not understand why she was expressing this to me. I didn't see the relationship because I was not aware of my rights. By Jessica bringing this up, it shows that she was poking around about discrimination, managing and accessing the potential risk of our dilemma. My lack of response and seeing the correlation let her know just how naïve I was.

EXPECTATIONS MEETING

Even after our last meeting, the same patterns of behavior resumed. We had a virtual meeting with team members to discuss the Fort Lauderdale bridge rehabilitation project, the one on which I was serving as project engineer. In the midst of a discussion about the results from a trunnion calculation and deciding on our next steps, the issue became evident. As already noted, this bridge is a bascule style, meaning it rotates open similar to a seesaw, making the trunnions critical components, serving as the points of rotation or fulcrums. They are large, slow rotating shafts taking the entire weight of the structure as the bridge operates.

During the call, I voiced my thoughts on some assumptions made about the structural loadings applied to the mechanical trunnion calculations. At the same moment, Mitch responded with a reassuring "Yes," while Nick snapped back with an angry

"NO!" My statement wasn't incorrect. Nick simply didn't take the time to listen to what I was saying. He seemed set on contradicting me at any opportunity he could, a clear autopiloted bias that was outright sexist discrimination. It was both unfair and exhausting to continually endure.

Following this meeting and Nick's biased treatment, I felt compelled to reach out to Mitch via instant message. Feeling uncomfortable about the prospect of facing Nick again, I requested a meeting to clear the air and establish some expectations. Mitch had already committed to scheduling a follow-up from our previous meeting, something he still hadn't done. The next day, Mitch responded, inquiring about what I hoped to achieve in the meeting. I provided him with a list of three expectations or boundaries of mine for discussion.

1. That we be professional and respectful.

2. Define tasks with, at minimum, expected or budgeted hours, as well as providing any previous samples and suggested manufacturers.

3. Work on unconscious bias.

Number one is obvious in context, I needed him to be more professional and respectful. The second expectation I set was in response to Nick's usual refusal to provide me with the budgeted hours, or at times much information at all on the tasks he assigned to me. I articulated this to Mitch, explaining, "He used to micromanage me. More recently, he has been giving me nothing." It was a sink-or-swim situation. Throughout our conversation, Mitch responded with jokes, attempting to lighten the mood, similar to what Jessica had done. The last point, undeniably, was about Nick's unconscious bias. I referred Mitch to the recent meeting which he and Nick instantaneously responded, "Yes" and "NO!" respectively.

The meeting to discuss expectations was scheduled virtually for the next day. We went through some items, but Mitch seemed to deliberately avoid discussing Nick's bias. This tactic, much like his jokes, again aimed to steer the discussion away from the topic of discrimination and prevent it from turning into a formal complaint. However, this time, I had managed to get some of my concerns in writing, which I kept a copy of.

Coming from Nick's perspective, he voiced that his pet peeves were not being heard and having to repeat himself. As it was brought to my attention in the meeting, I responded appreciative of the feedback and stated, now that I know I could try and work on that. However, he would refuse to listen or hear me and what I was saying. Later, in real time exchanges with Nick, it also helped me see that the problem was his aggressiveness, prior harassing, and how at times it made it difficult for me to focus and absorb what he was saying. In addition, he wanted me to blindly do as he said without questioning him by any means, however, it was literally part of my job to question things. That is the point of why we have the QC process, to check each other.

Mitch frequently shared anecdotal stories during discussions. In this meeting, however, he used these types of stories to try and shift the blame from Nick to me, attempting to portray my complaints of bias as an issue of my own bias. He stated that I was assuming Nick had bad intentions. I clarified that I couldn't assume Nick's intentions were good, given his words, actions, and the difficulties they caused. The numerous contradictions made it hard to believe in his good intentions. "Assume good intentions" I was told. Clear and intentional gaslighting.

Even during this call, Nick couldn't resist making comments that implied a lack of faith in my abilities and my capacity to complete the work in a reasonable time. Despite already being highly efficient, I was pushed to work harder and harder in increasing efforts to protect my job.

JUST KEEP TRYING

What else can you do? I enjoyed working for HDR, and I made that clear numerous times. I did not want to leave HDR. So, in addition to working harder and faster on my assigned tasks, I also began putting in effort outside of the movable bridge group. I searched for strategies to better manage my interactions with Nick and handle the ongoing issues. I proactively enrolled in several company management training courses on topics like unconscious bias, stress management, and navigating difficult interactions. Additionally, I took training courses including the Clifton Strengths as well as People Styles at Work the company offered. These courses gave me an opportunity to learn a lot about myself, but in retrospect it gave the company a great deal of information on me too.

Taking the Clifton Strengths assessment proved to be an eye-opening experience for me. If you haven't done it yet, I highly recommend it. Much like other personality type tests, it evaluates your strengths and then ranks them in order from your strongest to weakest, totaling 34 various strengths. My top five strengths, starting with the highest, are: restorative, responsibility, relator, learner, and analytical.

I take great pride in these strengths as they accurately drive and motivate me. The restorative trait reflects my enjoyment and skill in fixing things, which is a crucial part of what makes me a good engineer, especially when it comes to working on rehabilitations. I excel at problem-solving by identifying issues and resolving them. Responsibility indicates my dedication to my work, showcasing my honesty and loyalty. Relator signifies my enjoyment in forming close relationships with others, particularly when working toward a mutual goal, highlighting my value as a team member. Learner and analytical round out my top five. These traits align well with my engineering career—I have a passion for learning and growth, and movable bridges offer a non-standard work environment full of challenges. Lastly, my analytical nature drives me to understand cause and effect in various situations. I work to contemplate all factors involved in a problem, considering both the broader, global context and the finer details.

Along with discovering my greatest strengths, the Clifton Strengths course also opened my eyes to the diversity of people's abilities and perspectives. To me, my top five strengths make perfect sense and clearly represent how I operate. However, others have a completely different set of top five strengths. Many of us tend to assume that everyone thinks the same way and should react the same way. But the truth is, we are all very different. With 34 different strengths, there are 33,390,720 permutations or versions of how just the top five strengths can be ordered for an individual. If you take into account the order of all 34 strengths, that comes to a staggering estimated 2.95×10^{38}, or a total of 295,232,7 99,039,604,140,847,618,609,643,520,000,000 different per-mutations in which the order of a person's strengths can be listed from strongest to weakest. Wow!

As I noted, besides Clifton Strengths, I also took the People Styles at Work course. This was the only course that HDR had actually assigned to me, but Mitch and Nick were also sup-posed to take it. Unfortunately, they did not take it seriously. Before the start of the course, participants were supposed to take an assessment, similar to other personality type tests. The test results yielded your primary and secondary people styles, this time with only four styles to choose from: driver, analytical, amiable, and expressive. Drivers are the team motivators who drive work forward; they are strong, deci-sive, and results-oriented. Analytical individuals think before they act, dealing with facts, data, and logic. Amiable people are generally the most friendly, and expressives tend to react impulsively. My results indicated that I am an analytical driver, meaning my primary style is driver and my secondary style is analytical. While I personally believe these two might be reversed, these were the results of the test. Regardless of which is primary or secondary, these two styles confirm that I am highly task-oriented. I genuinely enjoy working.

Getting back to Nick and Mitch, only Mitch took the test. Nick told me he didn't even take it, so I never got to learn his style. The whole point of assigning the course to the three of us, from my understanding, was to learn how we could work better together. And then Nick didn't even take it. Mitch's results showed he was an amiable driver, which I think is accurate since he came off very friendly. If forced to, I would say Nick is an expressive driver. People with this style can have a short fuse and become loud and expressive easily. On the other hand, someone with both analytical and driver as their primary and secondary styles tends to not be very expressive until they are stressed.

Looking back, I'm not sure if it was all done intentionally. I was taking all these courses, and HDR was collecting data on my personality. Some of it was done by choice, and some of it felt like they were nudging and manipulating me to do it. At the same time, they could have assigned Nick several of these courses, but that would have left a trail of information on him as well. Even Nick not taking the People Styles course may have been intentional. They assigned it to him; he just didn't take the course.

SEVERAL HR DISCUSSIONS

I was doing everything I could think of to deal with what was happening at work, but it only got worse as time went on. There were several points that I reached out to discuss Nick's behavior with Jessica in HR. I was looking for advice or a knowledgeable and experienced perspective on this type of situation. While Jessica is very knowledgeable on the subject, that doesn't guarantee her good faith intensions on helping me. She was the one who originally suggested I bring up setting expectations for Nick and myself in the expectations meeting, which was good. However, she also spoke with Mitch after our discussion and before those meetings we had.

It's as if HR representatives try to herd you in a certain direction, treating us like sheep. She gave advice to me, then to them, and back to me, and so on. But was it all honest and ethical? Or was some of it deceptive? Manipulation seems to be a key part of their job. In one meeting, Jessica told me that I needed to "manage up" to Nick. Based on the discussion and my understanding, that meant I needed to call him out on things that he was not doing properly. Unsurprisingly, that didn't go well. I brought this back up to Jessica later on, and she backpedaled, trying to blame me. Anything to keep the liability off the company.

I turned to Jessica for guidance many times. I talked to her about Nick being dismissive and discounting towards me, and how he would do it in front of others which was disparaging. I said he was biased against me and treated the other men on the team with more respect. I felt like he was holding me back in my career, and I was unsure of his motives or intentions. He acted as if he was God and refused to give me credit when it was due. I told her he was undermining, disrespectful, and made me feel uncomfortable, small, and fearful. That he rarely gave me positive feedback and hearing it from others, like Mitch, had started to feel foreign. I even told her that it had made me start to not feel like myself, which I now recognize as a sign of the damage to my wellbeing.

COMPLAINING PROPERLY:
PUT IT IN WRITING AND FINISH THE SENTENCE

There is no doubt that Jessica knew what was going on. She was aware I was being harassed and discriminated against. That's what HR is trained for. After going through this entire ordeal, leading to consulting lawyers and conducting my own research, I realized that I was using all the key words and phrases to raise red flags on discrimination. However, I didn't fully understand what I was dealing with at the time. I knew it was wrong and I needed it to stop. I kept asking for help. As if drowning and all anyone was doing was barely enough to keep me afloat. They should have tried to pull me out, but they didn't.

If you don't put it in writing, they can try to deny it. I couldn't comprehend why, after so many complaints, so little was being done. I had too much trust in Jessica, thinking she was helping me. However, she was NOT my friend. She was not my ally. I was looking to protect myself. Her concerns were protecting the company and managing the risk of this all turning into a lawsuit. Thankfully, I was taking notes and I still have those notes. Those notes are documentation of what was happening along the way and act as evidence. For me it was self-preservation, but ultimately, I was preserving and documenting the facts.

Something I could have done better though, was to emphasize more that this was happening because of my gender. I had to finish the sentence and include "why" it was happening. Although Jessica was trained to see it, she wasn't asking those questions.

5

RETALIATING, CONDITION OF EMPLOYMENT, AND NEGLIGENCE

RETALIATION

What is retaliation? Why is it so wrong? One could take longer than a few minutes to think about it. Retaliation in simple terms is defined as a counterattack. You did something that upset another enough for them to intentionally come back and harm you. It's revenge. If we were to allow for this in our society, when would it stop? If people continued to counter-attack back and forth, would it ever end?

What if retaliation happens in the workplace? Well, it happens all the time, much more than we are aware of and more than corrupt corporations breeding toxic workplaces are willing to admit. Let's say an employee makes a complaint about their supervisor to HR. The HR representative will likely go back to the supervisor and discuss the matter. Instead of things getting better though, as you would hope, they often get much worse. The supervisor is now upset because you made their behavior known—essentially, got them in trouble. HR is also not happy because they will be on high alert, planning on how to minimize the company's risks of being sued, without genuine concern for the individuals involved. The result ends up being that the harassing supervisor is entitled to the illegal behavior with the support of the company behind them. Even worse, sometimes the supervisor will get a promotion so as to make them look like a reputable employee which transforms that entitlement to empowerment leading to the worst kind of upper management.

I had been on the receiving end of retaliation for a long time already. When I would complain to Nick directly about his behaviors, I would receive increased aggression from him. Even when I wasn't verbally responding, as with the sexual harassment, he would get angry and treat me worse because of it. Once I started complaining to upper management and HR, things got much worse and the retaliation became more intense. When you are unaware of these systematic tactics, it is very hard to see what is happening. It puts a person in a position where, eventually, they can just break down. There is also the option of leaving the company, which is what they could be trying to force you do, but I didn't want to leave. I loved my work and I believed in HDR.

How is it against the law? Well, it falls under its own category of retaliation, which is another form of discrimination. If an employee makes a complaint, specifically a protected complaint of unlawful discriminatory conduct, it is against employment and labor laws to then retaliate against the individual who is complaining, even if you don't use legal terms in your complaints. This protected complaint could be about any one of the various types of discrimination, such as gender discrimination or harassment, allowing for Federal protection. The retaliation could manifest as conscious or unconscious acts aimed at making your job difficult, being non-responsive, taking work away, disparagement, intimidation, threats, giving a poor performance review, or even termination, among many other things. Anything that would deter the complainant from making complaints.

CORRUPT PATRIARCHY, CONSPIRING, AND MOBBING

In some cases, like mine, the supervisor may bring in others to reinforce his retaliation. The term "patriarchy" has been used, often in reference to a male-dominated hierarchy. Yes, a patriarchy is a system or organized group where those in power are predominantly men. However, this fact alone does not make them oppressive or sexist; there needs to be more than just their gender involved. Nick, Laurence, and Mitch constituted a patriarchal order, being three upper-management males in positions above me. However, what they became was a *corrupt* patriarchy. If they are simply a group of men in authority, that in itself is fine. But if they conspire to engage in discrimination, harassment, and retaliation, that is not okay and is, in fact, illegal.

Nick and I were experiencing conflicts due to his sexist and biased behavior. Then, Laurence and Mitch began to protect him. The three of them worked against me in retaliation for my complaints about Nick's unfair and unlawful conduct. What they did to me was utterly wrong. The gradual progression of their actions made it difficult to recognize in the moment, but my documentation paints a clear picture of their intentions. Mitch once said something along the lines of, "I'm just going to keep an open mind throughout this and see what happens." I didn't fully grasp his meaning at the time. I couldn't see behind closed doors, but it seems they were conspiring, and he was attempting to remain impartial—at least he was in the beginning.

When people here the term "conspire" or "conspiracy", they tend to relate it to paranoia or make amusement out of it. I myself used to have the same association. Often when you hear something on the news about a "conspiracy theory" it tends to follow with jokes regarding the actions, discounting its validity. It is another misconception. The term conspiracy actually defines an agreement made, whether formal or not, by two or more parties to act unlawfully. The term "collude" is frequently used instead of conspire but, in reality, the latter is just the legal term for it.

But was the company also somehow involved in this? Another term sometimes used to describe such more specific situations is "mobbing." This occurs when the situation escalates even further. It becomes more like a crowd of people all attacking you, intended to instill fear, make you feel crazy, and drive you away. At this point, it was mainly the three of them, with Jessica also being involved indirectly or behind the scenes. However, later many others would join in, contributing to a mobbing I was subjected to.

MANIPULATION AND GASLIGHTING

Despite everything, I continued to be focused on my career. However, now that I started noticing these slight instances arise with Mitch and Laurence as well, it again added more stress. I did not have problems with them before this. It was because I was complaining about Nick's treatment that they began retaliating. They were not only seeking revenge, but also making up ways to create a trail of documentation against me. They would later try to use this to get me fired.

One example, occurred during a group virtual meeting about the Fort Lauderdale bridge. The main objective of the meeting was to discuss the initial scoping report we were preparing to submit to the client. This type of report is designed to document our field findings and provide a general assessment of the bridge. Additionally, this is the stage where we offer the client our recommendations, sometimes including alternative options, along with a budgetary cost estimate for the proposed work. Although I do not fully recall the exact details of the discussion, these details, along with details of other virtual meetings, should be retrievable, if necessary, after a lawsuit is filed during the discovery phase.

During the meeting, I believe Mitch veered off topic, bringing up a tracking spreadsheet that he had asked me to work on for him. This spreadsheet is a tool he, as the project manager, prefers to use to keep track of the scope items promised to the client within our contracts, as well as to monitor the budget spent to date. My role as the project engineer involved tracking the work performed and providing updates on spending to the project manager. However, at this point in the project, we had not yet submitted the scoping report, nor had the client agreed on our recommendations or finalized the budget for the job. It was premature and not a priority to spend much time on this tracking spreadsheet at this stage of the project. Nevertheless, I had begun working on it and had already informed Mitch of my progress.

During the call, however, he made a statement expressing a lack of "trust" in my having completed the task. This was the same person who had previously stated that he seeks out team members like me because I would keep him "safe". Suddenly, he didn't trust me? It was retaliation, and it was done during a meeting with other team members present. In my complaints about Nick, I stated that I didn't trust him. I said this due to his repeated unfair and degrading mistreatment towards me. Now, they were taking my complaint and turning it against me in manipulation. Similar to our Miranda rights, "Anything you say can and will be used against you". I had never realized how literal this statement could be.

I attempted to confront Mitch through instant messaging afterward, but he refused to acknowledge his words and actions, denying what he had said and subsequently shifting the blame onto me. He was gaslighting me, a specific form of manipulation and psychological abuse intended to confuse the victim and distort or alter their perception of reality. At that time, I still trusted Mitch and viewed him as a good guy trying to support me through dealing with Nick's unlawful conduct. I was wrong.

While Mitch began to manipulate and gaslight me, Laurence started ignoring me. He used to respond promptly to my emails and other forms of communication, but suddenly, he stopped replying to my messages, answering my texts, and returning my calls. At the time, I assumed he was simply busy, so I let it go and did my best to work around it. Little did I realize, both Mitch and Laurence were now also intentionally contributing to making my job more difficult, I just hadn't figured out why yet. They were actively working against me to protect Nick.

One day, feeling fed up with something Nick had done, I sent Mitch an instant message, exclaiming that Nick was lying. Mitch took that message, copied it into an email, and sent it back to me. Following that, I called him to discuss the matter. I explained to Mitch that I had written that message because there had been several incidents where it seemed like Nick was lying to me and attempting to mislead me, causing mistakes in my work. This is characteristic retaliation. At this point, I was still unaware of my rights or the concept of workplace retaliation. I was merely voicing my complaints about Nick's behavior which was negatively impacting my productivity. However, that was precisely the point. Nick was deliberately trying to force me into making errors and prolonging the time needed to complete my tasks. They were monitoring this through the Group Performance Evaluation, and he was attempting to skew the results.

What made matters worse was Mitch's response to my complaints. I provided Mitch with several specific examples of instances where Nick was being deceitful and misleading. These included Nick sending me an email specifying the number of hours allotted for a task, but then verbally telling me I had more time. Another example involved him instructing me to design a component for a rehabilitation project in a way that would interfere with existing bridge bracing. It wasn't until Nick went on vacation that I had the clarity of mind to realize I needed to redo the design. Mitch's advice to me was to stop complaining, asserting that I couldn't prove the deceit and told me that this was precisely why gaslighting is such an easy tactic. Although I had vaguely heard of gaslighting before, I didn't fully understand what he meant. He explained that lies are intentional, but it is impossible to prove whether someone is lying or they made a mistake, as they could simply deny lying. This entire conversation was yet another way I was manipulated, discouraged from complaining further, and my complaints not taken seriously. I had every right to voice my concerns and document these instances, as they were forms of retaliation.

APPLYING FOR SENIOR ROLE

Instead of deterring me, these events made me feel like I had to work harder, do even more, and achieve more. So that's what I did. I noticed that a position for a senior level mechanical engineer in our movable bridge group had opened up. Curious, I decided to inquire about it but was immediately blown off by Nick, receiving hardly any response. His reply was a curt "No," followed by a chuckle, as if the whole matter was a joke. However, I failed to find the humor in it.

I believed in my abilities and felt confident that I could take on the role. Nevertheless, Nick's constant belittlement had taken a toll on my self-esteem. More than that, I was growing concerned about how the disparagement from both Nick, and now Mitch, might be affecting others' confidence in me. Managers are supposed to foster career progression, not hinder it. But was I actually qualified? Were there specific requirements I needed to meet? Nick was not forthcoming with this information. Whenever I pressed him on why he wouldn't consider me for the senior position, his responses were dismissive, ranging from "because I said so" to "when the time comes."

However, this is not how things should be done. When an employee expresses interest in a promotion, they should be provided with clear guidance on how to reach their career goals, rather than being held back or, even worse, facing discrimination.

Having already reached out to others, such as Jessica, I decided to consult our office manager, Noreen Ryan, for more information about the senior level position and to discuss other topics related to career development. Following Mitch's unsettling comment about not trusting me in the meeting, I walked into Noreen's office, ready to present myself as someone eager to step up. We talked about the senior position, and she encouraged me to apply. She also recommended my joining the Women's Employee Network Group (ENG), which I did based on her advice. Furthermore, Noreen mentioned HDR's mentoring program, suggesting I sign up. When I eventually enrolled and received my mentor assignment, I discovered that Noreen was my assigned mentor. Was this merely a coincidence or not?

This time, I actually documented the discussion in an email. I sent her a list of the topics we had discussed, along with updates on the initiatives I had already started. She had also recently hosted a virtual office meeting, during which she asked for volunteers to deliver short presentations at one of our upcoming meetings. Following the meeting, I sent her the email and took the opportunity to express my willingness to volunteer for a presentation on movable bridges. She appreciated my enthusiasm and promptly scheduled me for one of the sessions. Unfortunately, my presentation was later repeatedly canceled and rescheduled due to various reasons by Noreen. I can't help but wonder now, was this somehow connected to Nick's discrimination and the retaliation against me? Regardless, I went ahead and created the presentation, choosing to deliver it at a Toastmasters meeting as a form of preparation.

Shortly after my discussion and email exchange with Noreen, I submitted my application for the senior mechanical role. I am still waiting for that promotion, because they revoked it after I blew a whistle. But that part of the story is about a year from this point. There is still a considerable amount of unethical and illegal activity to uncover between this time and then.

THEY TAKE ME OFF BRIDGE INSPECTIONS

A few weeks later, during one of our weekly progress meetings, it was my turn to provide an update. I reported that Laurence and I were planning to go out on inspection soon, but I had not yet received a response from him. As I mentioned before, he had been ignoring me. Nick then said that he wanted to talk to me about the situation after lunch, which I found odd. I had been conducting these inspections for years already; what could he want to discuss?

When we reconvened after lunch, Nick informed me that I would be removed from the inspection duties on this contract and that he would be taking them over instead. My immediate reactions were "What?" and "Why?" Moments like this I could see that Nick took pleasure in his actions. This conversation took place in October of 2021, and I was only given unjustified excuses for my being removed from the inspection duties. I immediately pointed this out to Nick as soon as he informed me, but he barely hesitated and stood by his decision.

Nick and Laurence were actively retaliating against me here and taking work away from me. But what was their ultimate goal? Were they just retaliating because they were upset about my complaints? That was part of it. Was it an abuse of power, with two 'good old boys' wanting to spend time together? Perhaps, though that was a secondary benefit for them. Was Nick questioning my abilities as a mechanical engineer due to an assumption that I was not doing a good job in my inspections? Yes, he could not see me as a qualified mechanical engineer, but rather, he saw me as an incapable woman. Did this also originate from his sexual intension towards me and my rejecting and ignoring those advances? I believe so. But I believe their primary motive around this time for removing me from the inspections was to further attempt to create a record of poor performance on my part. Nick intended to revisit some of my previous inspections from the year before and portray them in a negative light. What they did not consider, however, was that neither of them possessed the required training according to Federal guidelines to perform the inspections alone.

Another motive for, or at least outcome of, excluding me from these inspections was to hinder my career progression by limiting my exposure and experience. Being barred from participating in these inspections meant missing out on opportunities to hone my skills and gain firsthand experience with different bridges. However, they still expected me to perform the task of inputting their inspection findings and generating the reports, relegating me to more "secretarial" type duties. I had previously been managing this responsibility for Laurence on the electrical components which I was happy to do, but now I was being asked to do it for the mechanical aspects as well. This was yet another example of both discrimination and retaliation.

MORE DISCRIMINATION

During the very next weekly progress meeting, Nick followed his usual routine of asking each team member about their progress on tasks and their projected workload, leaving me for last. All of my male colleagues were afforded the opportunity to speak and provide their updates, but when it was finally my turn, Nick interrupted, dismissed me, and proceeded to input my hours into the progress spreadsheet based on his own assumptions. This disrespectful behavior exhibited blatant differential treatment; all the men were treated with respect and given a chance to speak, but not the woman. Not on his watch. Such actions from a supervisor could only serve to undermine my standing within the team. Nick was making my job difficult from every direction—be it through Laurence and Mitch from above, or by eroding my credibility with the rest of the group.

Two days later, I was back complaining about it to Jessica. I made yet another list of instances related to Nick's disparate, dismissive, undermining, and disrespectful treatment towards me along with how it was having a serious effect on me. I mentioned that I didn't have issues with others, and how an email response sent by Mitch the day before simply noting "Thanks Diana" alarmingly felt foreign due to Nick continual mistreatment. I also told her I had been taken off the bridge inspection task and that it was not okay. I still have these notes. Once more, documentation is very important.

A month later, we had a virtual meeting to discuss another contract. This project involved conducting a peer review for a client. A separate consultant from the movable bridge industry had completed a new design for a bridge replacement, and the client hired us to review their plans, calculations, and specifications. This task was substantial, given the extent of work involved in a full bridge replacement. We divided the design review responsibilities among the team. Nick and I took on different portions of the mechanical components, and then cross-checked each other's work. Other team members handled the structural and electrical sections, and I also contributed there to some extent. Once we had completed all the reviewing, checking, and organizing, we scheduled a call with the designing consultant and the client.

Aware of Nick's tendency to be dismissive and disparaging towards me, I decided to remain quiet during the call. Fortunately, everything went smoothly. Towards the end of the call after the others had left, only the HDR team remained, including Nick, Mitch, Laurence, myself, and one structural engineer. The discussion then turned to the specifications on the project. I now chimed in to point out that the specifications documents did not clearly establish precedence among the specification sections, which is typically stated explicitly to prevent questions arising during construction from discrepancies. My observation was both valid and pertinent.

Shortly after, Nick chose to make derogatory remarks about individuals with dual degrees, particularly in mechanical and structural engineering, claiming they "don't make good engineers." His comments were a barefaced dig at me and were both unnecessary and attacking. This instance was yet another example of Nick's inability to see or admit to my value on the team. When I attempted to contribute meaning-fully to a conversation or meeting, he would belittle both my input and my person. This constant demeaning was not only exhausting but also deeply hurtful.

Then, a few days later, I reached out to Laurence via email to discuss some of his electrical inspection notes from the first inspection that Nick had attended in my stead after I was removed from the task. I had noticed an inconsistency in one of his photos that was not documented in the pre-vious report or in his field notes. I asked Laurence whether he wanted me to update the notes or if there was perhaps something I had overlooked. The issue I was addressing con-cerned the installed orientation of one of the lights on the fender system, which protects the bridge from vessel impact through the navigational channel. These lights serve as navi-gational signals, indicating when it is safe for boaters lacking clearance to pass under the bridge, functioning similarly to a roadway traffic light for larger vessels. Consequently, the ori-entation of these lights is crucial for both boater safety and the safety of bridge users, necessitating adherence to specific orientation guidelines.

Laurence's initial response was that the installation was not 100% correct, but he was fine with how it was. Unsure if my previous relay of the information was adequate, I sent him a zoomed-out screenshot of an aerial view to better illustrate the issue, wanting to ensure he was fully aware of the situation. I told him that if he still felt comfortable with it after seeing the image, I would let it go, but I wanted to make sure he was fully aware especially since it was a safety concern. He then replied that he had been aware of the issue for the last seven years and was still fine with it. However, just because something has been a certain way for seven years does not mean it is correct nor should it be ignored. Complacency, especially in matters of safety, is dangerous and should be actively avoided.

In the past, Laurence had always welcomed and appreciated my diligence in pointing out potential issues in his inspection notes. This time, however, it almost seemed he was annoyed by my inquiry. The slight shift in his attitude caught my attention, particularly since I had already been complaining about Nick, although it was only starting. The change was occurring right before my eyes, but it was not easy to recognize at the time. Retaliation, like other forms of discrimination, are not always overt.

The very next day, I found myself being yelled at and talked down to by Nick during a phone call. I had sent him the print out of the report from the inspection him and Laurence went on and was expecting his review. I called him to follow up on the status of that so the report could be finalized. The conversation ended up shifting to him sharing details about a current issue with a contractor in the field. A contractor was experiencing difficulties while installing a rack gear as part of a construction project, originating from a design our group had previously created for a railway company. A rack gear is essentially a very large open gear, typically so large that it is designed in segments. In this case, the contractor was replacing the existing rack on a swing span style bridge, but the new rack segments were not aligning properly due to an incorrect radius noted on the plans. This was a significant problem, and Nick was understandably upset. In an attempt to calm him down, I said the contractor should have field verified the dimensions. His response was yet again aggressive; he yelled out, "You don't know what you are talking about!" This was yet another instance of degrading and insulting harassment, as he consistently viewed and treated me through a gendered lens, exclaiming me not capable.

My intention with the statement was merely to point out that the responsibility for field verification lay with the contractor. I was implying that the error was not on our end, as I knew Nick tends to react negatively to mistakes. Unfortunately, he chose to respond with hostility. This was again uncalled for, and it's exactly the type of comment that can diminish a person's confidence. To add to this, about a year later, we had a virtual meeting with the same contractor about a different bridge for the same client. The project was similar, and this time the contractor requested dimensions in advance to ensure they could field verify them. They even referenced the previous job during the meeting, expressing their desire to avoid a repeat of the previous issues. This affirmed that my initial statement had not been wrong.

To clarify, my issue here isn't about being right; it's about Nick's aggression being completely uncalled for. Enduring this type of conduct for two and a half years was causing me harm.

"WE'RE GONNA GET RID OF YOU"

What would you consider "severe" harassment? Has any of Nick's conduct that I've described so far seemed severe to you? Think about your own experiences. Have you ever been subjected to severe harassment? What is it that renders the conduct or behavior severe? Is it the words spoken to you, or the actions taken against you? Or is it how you perceive those words and actions? How does it affect the individual who is being harassed?

Now, consider the term "pervasive." How long must harassment persist before it is deemed pervasive? This concept is not clearly defined. According to the EEOC website, whether harassment is severe or pervasive enough to be illegal is determined on a case-by-case basis.

For me, this next incident was severe. It demonstrates malice. I walked into Nick's office one day, by this time he had moved into Laurence's old office, as Laurence was frequently out of town for work. I was in search of additional assignments since my workload was light. We discussed the remaining assigned tasks I had and the various reasons I couldn't progress with them—either the project number for charging time wasn't set up yet, or we were awaiting client responses to proceed. Out of nowhere, Nick threw in, "We're gonna get rid of you." This wasn't the first time he had threatened my job; he previously threatened it after I CC'd Mitch in an email. And now this!? More importantly, I didn't recognize until better understanding the law, this comment indicated overt intensions between him and at least one other to act illegally – they conspired. In the moment though, I managed through the conversation by generally ignoring the comment and only expelling a nervous laugh. But as soon as I left his office, fear and anxiety over- whelmed me. The instances of him threatening my job were severe causing stress levels to soar. The other comment that had a severe effect on me this far, was "no one cares about you." Only, that comment didn't create anxiety but, instead created feelings of depression. In addition to these verbal statements, there was also the sexual harassment and instance of sexual assault that occurred which was severe, even though my mind was endeavoring to ignore and compartmentalize that.

There I was, attempting to extract more information from Nick to determine which task I should prioritize, demonstrat- ing a responsible and diligent approach to my work. And there he was, insinuating that it didn't matter because I was on the verge of being fired. Still, I maintained my professionalism, focusing on the work at hand. The mantra "just focus on the work" again repeated in my head, as it often did in coping with Nick's behavior. At the end of our discussion, he said he would come to my desk to talk more in a little while. When he finally did, it mainly involved him scrutinizing my work and nitpicking, all while attempting to belittle me within earshot of our colleagues in the office.

I was on the receiving end of unlawful conduct which went much deeper than I knew and it was affecting me in a serious way. I had already complained about it several times to several people. Yet it continued to happen. Something had to change, and I was ready now to explicitly request it.

CONDITION OF EMPLOYMENT

It was now the beginning of December 2021, and I found myself at another critical juncture. Nick and Laurence had recently removed me from some bridge inspections, and now, I was told that they were trying to "get rid of me." I knew I had to speak up loudly and clearly this time. It was obvious that Nick was not on my side. I had to protect myself.

I reached out to Mitch again, after my employment was further threatened. I told him what had happened. Mitch quite instantly tried to deny it, although he wasn't even there! How could he know what Nick had said? He was gaslighting. I very specifically told him I needed a new supervisor, as I couldn't work under Nick any longer. I also asked him questions about how he viewed me and my future at the company, but he didn't give me a direct answer. With regards to the new supervisor, his response was to check with HR.

When someone threatens your job, it can throw you into a tailspin of doubts being concerned about employment, career, and livelihood. Finally, a few days later, we had a scheduled virtual call with HR to discuss the matter. On the call were Jessica, Mitch, and myself, although Mitch did not say much. I officially requested a new supervisor. I stated that it had been over a year since my first complaint to Laurence while out in Virginia. I had made numerous complaints again within this time. I mentioned several specific instances of harassment from Nick, including him saying, "we're gonna get rid of you," "No one cares about you," "Sit the f@#k down," and "You don't know what you're talking about." While the statements I brought up were not facially or overtly specific to my being a female, they are still covert harassment and discrimination, it's just not explicit. Had I understood my rights I definitely would have brought up a lot of other instances that were more explicit as well as the sexual harassment. In time, I do. For now, I also brought up his broad insulting and degrading me with other negative put-downs and how he was threatening my job again. I made it clear that this had caused me to lose trust in him and that I felt uncomfortable, nervous, and fearful. It was even affecting my health.

Jessica asked me questions about the specific facts surrounding Nick's statement, "We're gonna get rid of you." She wanted to know his exact words and the context of his statement to gather all the necessary information—the five W's: who, what, where, when, and why. It was very easy for me to offer up the who, what, where, and when, however, I hesitated when asked with regards to why. Within this, she also probed to determine the responsibility of the actions and the requests as they unfolded. She asked questions like, "Who made the decision on what to work on?" and "Why did Nick come out to your desk afterward; did you ask him to?" I did not ask him to. Even then I responded in good faith stating that she was asking for facts but I could only provide my opinion. Although the law apparently relies on the why, the why can be subjective, leaving literally the health of an employee being exposed to illegal misconduct, as I was, up to what employers make into a game of subjectivity. Jessica also noted that she would need to follow up with Nick to get his side of the story.

Still in that moment, I continued to seek a better understanding of how to handle the situation. Why or how could it be acceptable for him to make such comments to me? These were not jokes; the comments were personally directed at me. I wanted to understand, as I was unaware of how to protect myself. I also highlighted the fact that a demeaning comment could be just as, if not more, hurtful than the use of explicit language. Jessica responded that it depended on the comment and the manner in which it was delivered. She also raised questions about my perception of Nick's comments during our discussion. However, this wasn't a matter of perception; it was a direct threat to my employment.

Still, I tried to find a middle ground. I expressed my willingness to work with Nick if absolutely necessary, but stated my need to be placed under a new supervisor. I had an extremely difficult time thinking and concentrating around him. Jessica flatly denied this request, stating there was no option for a change in supervision. So, I pressed on, emphasizing, "But I feel Nick is causing me harm." Her response? That we would need to have a series of discussions. Discussions? I had just expressed that I was experiencing harm!

Confusion set in at this point. How was it possible that despite expressing harm being caused, Jessica still wasn't agreeing to assign me a new supervisor? How could any of this be acceptable? I pressed on, explaining, "It is very stressful working with Nick." That's when she suggested options for other positions outside of our area. Here I was, expressing my distress about a boss who was instilling fear, stress, and harm in me, and she was suggesting I should relocate? Was she seriously implying that I needed to upend my entire life because of his offensive, abusive, and threatening behavior? This was beyond comprehension.

Reflecting back on that period of my employment with HDR, it is very upsetting. It's clear how little my wellbeing mattered. The negative effects of working under Nick's supervision were having a damaging impact on my mental and physical health, and Jessica was not taking the necessary action to prevent it. The laws and company policies are set up in a way that an employee can be considered in a supervisory role over another, even if they are not the direct supervisor. To them, changing my direct supervisor wouldn't have made a difference if my workload still came from Nick. I did not know this at the time.

Recalling from Chapter 2, where I first discussed harassment, there are two scenarios in which harassment becomes unlawful. The first scenario was when harassment becomes a condition of employment, which is exactly what Jessica did by forcing me to continue working under Nick's supervision.

At the time, I couldn't see how any of this was justifiable. Now, I realize that one thing I hadn't plainly stated in the meeting was that this harassment was happening to me because I was a woman. I was aware of it, they were aware of it, yet I hadn't voiced it outright. But should I have had to? The employer, specifically HR, is responsible for ensuring compliance with labor laws and providing just and favorable work conditions for their employees. They are supposed to protect us from harm and ensure our safety within the workplace. It wasn't my job to be well-versed in employment and labor laws; I was hired to provide engineering services, which had become nearly unbearable due to the discriminatory actions of my supervisor. Nevertheless, as a woman, I belonged to a protected class, and Jessica was fully aware of it.

Even though I hadn't yet openly stated that the harassment was related to my gender, they were aware that it was because I am a woman. However, they failed to take appropriate action. The next issue I brought up was personal space, recounting times when Nick would invade mine. I described how he would stand too close to me at my desk, stating "sometimes he is right on top of me", making me uncomfortable. These were the moments when he would press his groin against the back of my chair, a detail I found myself unable to articulate fully. Jessica offered some basic suggestions on how I could address the situation, such as asking Nick if he needed me to move so he could see the monitor better. However, this advice didn't address the underlying problem, and speaking up to Nick had most times been a challenge for me. His actions, over my now two and a half years working for him, had left me feeling intimidated and stifled, afraid of how he might react if I expressed my discomfort.

The whole situation was unacceptable. The issues were evident, and I didn't know how much more explicit I needed to be for them to understand. Jessica was allowing this to continue, even after I voiced my concerns and explored potential solutions. At times, she would even redirect or shut down the conversation. When I asked what would happen if things didn't improve, she noted she would still refuse to assign me a new supervisor. It was entirely disheartening.

As an HR representative, Jessica was doing her job of protecting the company, but she was failing miserably at her duty to protect me.

NO CHANGE HERE, NO CHANGE THERE...EQUALS NEGLECT

Realizing that Jessica was firm in her decision not to assign me a new supervisor, I contemplated other options. What if I began speaking up to Nick more assertively? Could this earn his respect? I recalled Jessica's earlier suggestions about managing up. Reflecting on the situation now, I see things with clearer eyes. This approach was doomed to fail. Nick's bias was too deeply ingrained. The way everything was being handled—or rather, not handled—was exacerbating the situation.

More crucially, on the occasions when I did try to stand my ground with Nick, my actions were later twisted and used against me, painting me as the problem. They began to try to label me the aggressor and unwilling to take direction from him. It's a tactic they used to create false narratives—a key point to bear in mind.

Neglect, too, is a form of abuse. It's the failure to provide proper care for something or someone. In this case, Jessica was neglecting her responsibility to ensure a safe work environment and HDR was failing in its duties as an employer.

It was almost the end of the year and with anticipation about the next inspections, I reached out to Laurence and Nick again informing them of my availability to perform the upcoming inspections. To my disappointment, Laurence responded that Nick would continue handling the inspections in my stead. Again, no changes were made, leaving me baffled. No change here, no change there.

Negligence can manifest in ways beyond failing to care for a living being; it can also pertain to failing to fulfill job responsibilities. There was no valid reason for preventing me from conducting inspections, yet there were good reasons why I should have been assigned to the task. It was a case of gross negligence and mismanagement.

THEY WERE AWARE

The meeting in which I asked Jessica for a change of supervisor took place in December. A few weeks later, in January, we spoke again during a follow-up scheduled by Jessica specifically as a phone call and not a virtual call, avoiding future evidence of the call. During our conversation, Jessica inquired whether I believed Nick's treatment towards me was because I was a woman. I affirmed with something along the lines of "I think so." I also have in my notes from the call, where I wrote that there was male dominated favoritism. Unfamiliar with my rights and unaware of how the law applied to my situation, I was apprehensive about the potential repercussions. I realize now that I shouldn't have been fearful; I should have confidently asserted my experience. Shortly after our call, Jessica emailed me the HDR Harassment Policy. She was fully aware of the gravity of the situation.

Let's recap the situation: Nick threatened me, which made me feel unsafe. I reported this to his supervisor and to HR, and in both instances, I requested a new supervisor—a request that was denied, making the harassment a condition of my employment. I articulated repeatedly and consistently that Nick had been treating me unfairly for an extended period—pervasive harassment. I conveyed that his behavior instilled fear in me and was causing me harm—severe harassment. Eventually, I acknowledged that this treatment was gender-based.

Jessica was acutely aware of what was transpiring. Knowing all that I had shared in the previous meeting, she deliberately chose a phone call for our follow-up discussion. Virtual meetings create records that can later be discovered, and a phone call avoided leaving such a trail. I was naively unaware of the situation unfolding around me. She, along with others, was taking full advantage of my lack of understanding. Looking back, I should have documented via email that I had identified the issue as gender-based—a matter relating to my status as a member of a protected class—and retained a copy. I cannot emphasize enough the importance of documentation in situations like these.

THE CONFERENCE ROOM: ABUSE OF POWER, ENABLING, AND MORE MANIPULATION AND GASLIGHTING

I still found myself grappling with Nick's oppressive management style and unjust tactics. He assigned me the task of reviewing a mechanical maintenance manual for a bridge belonging to one of our railway clients. A maintenance manual is created to provide guidance to maintenance personnel when they conduct upkeep and repairs on a bridge. This document is usually produced by the contractor upon completion of the construction work derived from a design project, but before the contract concludes. In our role as designers, our job was to review the manual.

First off, I sought confirmation from Nick regarding the budgeted hours allocated for reviewing the manual. He indicated that it was a 32-hour task. At this point, my interactions with Nick had left me perpetually on edge, apprehensive of potential ulterior motives and retaliatory actions designed to trap me into making mistakes. His prior assertion that they were plotting to "get rid of me" only fueled this. Aware of the 32-hour timeframe, I nonetheless worked diligently and swiftly, as I did with basically all my work for Nick, Mitch, or Laurence. It felt like a tense game of hot potato, and I was desperate to avoid being caught holding the work when the metaphorical music stopped, which I now believed could be used towards my termination. I completed my review of the manual, annotated my comments, and sent it back to Nick, noting that the task had taken me only 13 hours.

I should mention here that our usual practice includes a checking process to ensure the quality of our work. However, Nick, ever the "boss," did not perceive himself as fallible or capable of making mistakes. With this in mind, he reviewed my work, altered a few comments, and sent it directly to the client. This maintenance manual was no brief document; it spanned around four hundred pages. Despite completing the task in roughly a third of the allotted time, I had still managed to compile a substantial list of comments and feedback.

Always eager to build on and improve my skills, I was looking forward to receiving feedback from the review. Unlike Nick, I am well aware that I am capable of making mistakes or overlooking details. After almost a week of silence, I decided to reach out to him to inquire whether he had checked my review, so we could proceed with sending the document back to the client. Nick, still unwilling to delegate more to me, at times lagged behind, creating bottlenecks in our workflow. I frequently needed to send him reminders and check-ins regarding deadlines. He eventually responded, confirming that he had sent the document back to the client that morning.

Curious, I asked if he had made any changes and requested details on what they were. I also pointed out that I had a substantial number of unused hours from the allocated time for the task. He replied, listing some of his modifications. My name was stamped on the document as the reviewer; he should not have sent it back to the client after making changes without first allowing me to review his changes. Following my response to his message, he replied with a threat to take the work away, stating that he could call the manual back from the client and put his own name on it if I preferred. More threats to strip work away, particularly after one has made protected complaints can constitute retaliation and is illegal.

Upon examining the changes he made, I discovered a mistake. *Here we go*, I thought. This discovery led to a series of emails being exchanged, during which he adamantly refused to acknowledge his error, attempting instead to wrongly justify his actions. It was an overt abuse of power. This was not a matter of preference or a change of mind, as Randall might say. With Nick, his decisions were variable and sometimes impulsive, dependent on his mood that day. This mistake though was undeniable, and his refusal to admit and rectify it rendered him a liability. Nick was acting in retaliation and bad faith by abusing his power, and allowing such behavior to go unchecked would only perpetuate it.

I reached out to Jessica once again, forwarding her the email chain as an additional example of Nick's behavior. A follow-up meeting with Nick, her, and myself had already been scheduled. However, what took me by surprise was Jessica's actions and demeanor in the upcoming meeting. Unlike the previous virtual meeting with Mitch, Jessica opted for an in-person gathering in a conference room. Was this another deliberate choice to avoid leaving a digital record? I think it was.

Right from the outset of this meeting, Jessica controlled the conversation, shifting the blame onto me. She began to portray me as the problem, suggesting I had done something wrong, causing Nick frustration. One of the first things she mentioned was the email I forwarded and that the meeting's focus would shift in a different direction.

Several times throughout the meeting she targeted my perception, starting with this email and her saying "I don't think it says what you think it says." Was this a tactic? It seemed intentional, a ploy to disorient and manipulate me. Despite this, my commitment to the quality of our work and my determination to stand up for what is right remained unwavering. I explained it was possible she was not understanding the mechanical jargon.

We were dealing with a swing span bridge. At the center of such a bridge, you will find a circular track and what are referred to as balance wheels that ride along this track. This setup ensures the bridge remains in reasonable alignment while in operation. The balance wheel assemblies are mounted to the underside of the bridge, supported by a structure, while the track is anchored to the substructure pier. These balance wheel assemblies feature either one or two bearings on a pin, or shaft, facilitating smooth rotation with low friction. In the case of a single bearing, it is housed directly within the wheel, with the pin held stationary. Alternatively, with two bearings, the bearings are located within the balance wheel support, allowing both the wheel and pin to rotate in unison.

The mistake Nick was refusing to acknowledge was straightforward. Undeniably, 2 x 4 equals 8. This particular bridge featured four balance wheels, each containing two bearings within its assembly. Consequently, there were a total of eight balance wheel bearings. In my initial review, I corrected a quantity in a table, changing a 4 to an 8 to accurately represent the components requiring lubrication. Fundamentally, with a wheel assembly, you specifically want to lubricate the bearings, not the wheel itself. This was a simple oversight, yet Nick altered my correction from 8 back to 4 without investigating the type of wheel assembly in question. He subsequently refused to acknowledge his mistake, compromising the quality of our work for the client. Maintenance manuals like these are utilized in preparation of maintenance; they guide maintenance personnel, potentially without prior experience, on necessary procedures. However, Nick seemed indifferent to the purpose of our work and our responsibilities to the client, choosing instead to cover up his error and undermine my credibility.

I expended a great deal of effort explaining this to Jessica during our meeting, while she seemed to continue to circle the conversation. She claimed to grasp the mechanical aspects, but her understanding was lacking, and she continued to blindly support Nick's stance. She backed Nick acting in the interest of commanding his authority, regardless of the truth or what was right. Jessica overstepped the boundaries of her expertise by discussing technical mechanical terms, an area well outside the scope of human resources. In doing so, she enabled Nick's abuse of power.

Furthermore, Jessica's approach made me feel attacked. She persisted in attempts of shifting the blame my way, accusing me of not listening to Nick and lecturing me on the importance of respecting seniority. While I have no issue respecting those with more experience, my tolerance wanes when they act unfairly and unethically. She instructed me to ask more questions, stating that managing up is all about inquiry, and implying that I was misunderstanding. I tried explaining that Nick was often unwilling to acknowledge his mistakes, emphasizing that this was a clear-cut error he was aiming to conceal, not my misunderstanding. Despite my efforts, Jessica continued to dismiss my concerns, shut me down, and work to silence me, while ignoring Nick's outright abuse of power.

Jessica expressed uncertainty about why I was fixated on what she considered a minor detail. To me this was a clear matter of fact, no subjectivity here, Nick was using the situation to act both in bad faith and retaliate against me. She also suggested comparing other similar examples as a standard, and I agreed, recommending we look at the other components within the table to compare. She immediately shot down my suggestion. Jessica further asserted that I pick and choose my battles. I explained that I had precisely raised this issue because it was part of a larger pattern of behavior from Nick, involving deception and an abuse of power.

Nick argued that maintenance personnel would simply go to the lubrication manifold without paying any mind to reading labels which identify what they were lubricating. I quickly countered, pointing out that some components require lubrication at different intervals, and some could also require different types of lubrication, emphasizing the importance of the tables in the manual. They both repeatedly tried to downplay the issue, insisting it didn't matter, but that is not how business should be done. We should aim at delivering a quality product to our clients.

The worst part of this was that our conversation was supposed to revolve around his harassment. How was it that she was completely sidestepping the topic, especially after sending me the Harassment Policy just the previous week? I attempted to steer the conversation back to the initial purpose of the meeting. It was about midway through, I asked if we would return to the original topic, to which she curtly replied, "No, we are not." Subsequently, with intentions of interjecting more politically correct language to safeguard the company, she asked Nick a leading question, prompting him to affirm my abilities as a good engineer, which he did.

Frustrated that she was neglecting to address Nick's general harassing behavior, I decided to speak up and voice my complaints anyway. I mentioned my discomfort and difficulty even looking at Nick during this meeting, which Jessica did acknowledge. Even his presence often left me feeling intimidated and scared, impairing my focus. I voiced my general discomfort around him, attributed to his mistreatment of me. Once again, she attempted to prevent me from speaking on the topic. She highlighted Nick's patience and responsiveness during this meeting, though his demeanor was again uncharacteristic and seemed prepped. These were attempts of altering my perception and deliberate gaslighting.

I continued to present a list of grievances I had from prepared notes, including Nick's remark that "no one cares about you." Predictably, Jessica interrupted, demanding facts and prompting me to recall specifics—remember, the 5 W's. I provided the details without pause, and Nick initially denied recollection before conceding he had he believed he said "no one cares about what you think." A different statement to the one he had actually expressed, I recall because of how hurtful it was.

Jessica shifted again to protect the company, suggesting a misinterpretation on my part and implying my memory might be flawed. More manipulation and attacking my perception. Nick, aiming to downplay stated his comment was meant as "just a jab", as if a jab would be suitably professional. Jessica, once more shifting, quickly reframed Nick's response, insisting he was being "sarcastic". More gaslighting.

When prompted by Jessica for additional examples, I cited Nick's frequent dismissive remarks, such as "you don't know what you're talking about" or "you're confused," and his tendency to resort to degrading put-downs. I emphasized that Nick's tone and facial expressions added to his comments coming off as demeaning and that I didn't get that sense with others. Jessica asked, "Are you holding Nick to a higher standard than others?" I reiterated, "no", it was his tone and actions.

While I didn't need constant positivity, I couldn't tolerate repeated threats, negative put-downs or "jabs." She continued to promptly interject, labeling Nick's behavior as "sarcasm." In the HDR feedback training course, they categorized two types of feedback, redirective and reaffirming. I clarified that redirective feedback, when delivered respectfully is not a problem, but Nick's disrespect was a problem. We all have a right to a respectful workplace— it is a fundamental human right.

I then addressed the issue of Nick threatening to take away my work or even my job, citing the specific email where he wrote, "you will find yourself not working for me any longer, understood?" Jessica questioned Nick if he recalled the email, to which he responded that he did not. As she turned back to me requesting more information, I began to respond and she again abruptly interrupted, suggesting that I was the cause of Nick's frustration, cutting me off before I could answer her question.

She attempted to steer the conversation back to sarcasm, bringing up Mitch's sarcastic nature. I acknowledged that while Mitch may joke around, he was not insulting. Nick now interjected, claiming that Mitch had advised him to take work away from me—a statement Mitch had never communicated to me directly. It was so out of left field that it took me a second to take in. I wondered if Nick was attempting to create discord between Mitch and I. But now, looking back I wonder if this was part of their covering up. Back in October, only about three months prior, I mentioned in a discussion with Jessica that I didn't have problems with others and that even Mitch's "Thanks Diana" email response seemed foreign after working under Nick's treatment for so long. On top of that, just about a month prior in December, I suggested Mitch as one of the options for a new supervisor.

Despite voicing these concerns, they continued to manipulate and gaslight me, insisting that I was misperceiving the situation, which left me once more feeling defeated and drained. Nick seemed to take pleasure in the things he was able to get away with. Jessica's apparent shielding of Nick coupled with a sense of power by her dominating the conversation seemed to be enjoyable for both of them, while it was causing further harm to me. Nick witnessing her actions, only enabled and empowered him even more to continue his misconduct.

But I didn't give up, still wanting to address the issue of disparagement. I brought up issues I've had with clients due to Nick insulting me, yelling at me, and being disrespectful in front of them. I explained that it compromises the relationship and makes my work more difficult. I mentioned that some of his actions can be overt and obvious, like when he is yelling, but other times they are covert; the disrespect isn't clear, but the undertone is there. I expressed my concern that it might affect my potential to move up in the company.

Nick jumped in, attempting to blame me. As we went back and forth again, it was two against one with both of them blaming me. Again, it felt like we were going in circles, or you could say they threw me for a loop. I specified it was when Nick disparaged me out in Virginia that I called Laurence and made my first complaint. Nick then made further accusations which were untrue. The whole thing was exhausting. I explained that the relationship with the contractor would not have been compromised if Nick had never been out there. How could they respect me if my boss doesn't have respect for me? We also discussed the inspection contract which we had limited hours on and he had requested us to work without pay if needed. Nick tried to argue that I was removed from that task as well, however, that was due to the way the contract was setup and his disparagement. I was removed from the task in retaliation of my complaints.

Jessica trying to put an end to the meeting told me I would have to let go of the past with Nick to move forward. She told me never to mention anything that has happened prior to today's meeting ever again. None of this was okay, particularly since she had just sent me the Harassment Policy and then made great efforts to avoid a conversation on harassment. The meeting finally ended, and Jessica said she would set up a follow-up in four weeks, but that never happened. Beware of conference room meetings like this.

IT WAS INTENTIONAL

What HDR was doing was extremely intentional. They were, just as Nick had said, planning to "get rid of me," and Jessica enabling it. The majority of my experience at HDR was packed with manipulation. I saw it then, especially during the last meeting in the conference room, but I didn't acknowledge or understand what they were doing at the time. It was as if they were working from a playbook. While employers are well experienced in these situations, the employees who are suffering are unprepared and ill-advised to handle the circumstances they are faced with.

While they dismissed and downplayed my complaints, I thought they were not taking this seriously, but I wasn't conscious of the fact that they saw the situation for exactly what it was, taking it very seriously and actually working in bad faith to avoid liability and push me out. Even though Jessica had sent me the Harassment Policy, I *still* didn't understand it. If I didn't understand, I know there must be so many others who don't understand or know their rights. It is what allowed them to take such advantage over me as they did. It's like employers are playing a game, but the employees don't know the rules. It's like cheating!

So, let's recap again. In December, we had a virtual meeting where I complained and asked for a new supervisor because I could not take it anymore working under Nick, and he had now threatened my job more than once. The following month, Jessica set up a phone call and asked me if I thought this was due to my protected class as a woman. She then sent me the Harassment Policy. The next meeting was scheduled in a conference room, and Jessica tried to avoid a discussion altogether with regards to harassment and Nick's behaviors. She also told me not to mention anything Nick had done again.

Jessica intentionally set up that phone call and in-person conference room meeting to prevent a record of what would happen. She also intentionally made the conference room meeting with Nick, from the start, about blaming me for the email I forwarded to her. Jessica specifically did not mention the Harassment Policy during the meeting. She took a meeting that was supposed to be about Nick's harassment and unjust discrimination and turned it right around onto me. It was purposefully done this way in bad faith and negligence.

When I realized all this, it knocked me down all over again. Depression flooded in, and that intrusive thought "no one cares about you," which came from Nick, began to repeat in my head once more.

This type of thing must be happening to so many others. It can be enough to make someone want to hurt themselves—enough to make someone want to hurt others. The lack of care in addressing the problems caused by the abuser, and then instead blaming the victim, is so wrong.

HR is not there for the employees. It is a conflict of interest. This is a major issue that is plaguing our society. Corporations are causing harm to individuals by dismissing their complaints and downplaying what is happening to them. This is then followed by blaming, manipulating, and gaslighting the victims in attempts to make them question their reality and withdraw their complaints. It's applying pressure on the one who is being abused instead of on the abuser. Discrimination, harassment, and retaliation are illegal. Nonetheless, corrupt corporations will maliciously sacrifice the wellbeing of the individual employee who is complaining in order to hide the truth and conceal unlawful misconduct.

They knew what they were doing, and I was under stress and focused on getting the things Nick was doing to stop that I didn't realize it. Fear can actually make you more susceptible to manipulation. Is that something that is also in their playbook? Is that why Jessica used that email to put me on the defensive?

I see now more clearly what Jessica, as my HR representative, was doing and how Mitch and Nick were guided to protect the company and avoid liability. It appears there were six main manipulation and gaslighting tactics:

1. Keep the abuser calm and polite during meetings about complaints.

2. Interject politically correct language to redirect away from facially or inexplicit discrimination based on protected classes.

3. Minimize the blame on the abuser and downplay his misconduct.

4. Place blame on the victim by labeling them the cause of the problem and starting false narratives.

5. Ask leading questions and set traps against the victim within conversations.

6. Directly target the victim's perceptions as inaccurate as well as to alter their perceptions.

First, the most apparent way in which Nick was coached was to ensure he remained calm and polite during these meetings. It was so noticeable, as previously mentioned, I even directly called it out during one of the meetings, because it was precariously out of character. Additionally, it seemed as if Mitch and Nick had a plan on what they were going to say, how to run the meeting, what to ask, and how to respond. It's like putting two people on an obstacle course but letting one walk through it first; that individual will naturally have the advantage. However, when Jessica was present at the meetings, she would dominate conversations, controlling the narrative.

Next, what do I mean by politically correct language? It was their way of slipping comments into conversations to negate the notion that Nick had a sexist bias or was discriminating against me. Examples of politically correct language they used included statements like "Nick treats everyone the same" or "He answers everyone's questions in the same manner." Flat out lies. They would also prompt Nick to agree that he believed I am a competent and capable engineer, despite his normal speech and actions suggesting otherwise.

Third, there was specific downplaying of my complaints, an attempt to minimize Nick's actions and misconduct. When I said, "Nick is aggressive," they would describe him as "harsh" or "gruff." When I said, "he is threatening," they would claim "He is using sarcasm." My perception is valid, and when they did this, they were gaslighting me. Because I was unaware, I occasionally found myself adopting their language, which is exactly what they wanted. They even began to shift blame onto others, such as Mitch and Laurence. An example of this is when Nick claimed during the conference room meeting that Mitch was the one who told him to take work away from me. As Nick was the abuser, this worked to take some blame off him.

Although all these tactics were harmful, this one was particularly damaging at this time. They tried to label me as the problem, even though I was doing everything I could think of to fix or work through the problems with Nick. They would say things like, "Nick is acting that way because you are making him frustrated." They also told me that "you have to listen more and ask questions," or "you are too analytical." These were attacks against me and also the beginning of their false narrative that they continued to perpetuate. While it was happening, I began to lose myself. They worked to strip away my dignity and self-worth. Way later, during recovery, I had to work to undue this and remember who I am.

Fifth, posing leading questions and setting traps. When they did this, I instinctually could tell that something was off. The questions made me pause, resulting in long silences before I would answer. I could sense the bad faith but was still naïve and trusting. Questions like Jessica asking, "Are you holding Nick to a higher standard than others?" or Mitch asking if I agreed that communication with Nick followed a pattern: "You ask a question, Nick responds, you understand Nick, but you don't agree with him." Moreover, they would set up scenarios to mislead me into making mistakes, so they could report poor performance. I had pointed out several instances of this behavior from Nick to Mitch, but was told to stop because I couldn't prove it. These are all retaliatory traps, so one must be careful.

Last, but certainly not least, they can gaslight, directly targeting your perception or trying to alter it, forcing you to doubt yourself. This doubt may cause you to complain less because you'll start to question yourself more. Trusting yourself and knowing the truth is crucial. For example, at the start of the conference room meeting, Jessica spoke with regards to the email I sent her and said, "I don't think it says what you think it says." They also told me that I was "misinterpreting." Both of these directly target my perception as inaccurate. An example of them trying to alter my perception was Mitch saying "you are assuming bad intentions." I was not assuming; I had witnessed Nick's bad intentions repeatedly. They insisted Nick's intentions were good, and at the time, my response was that I wasn't sure they were.

It was all manipulation and gaslighting. This was having a tremendous effect on me and my health, as they recurrently introduced lies and deceit into my experiences. The magnitude of this is yet to be sufficiently addressed, and are all things to look out for as part of toxic employer's systematic approach to protecting the company.

As an employee, though, it was mentally draining to deal with and would even make me physically tired. Manipulation almost feels like a heavy invisible blanket being pushed onto you and your mind. You can practically feel the blame and contradiction being laid upon you like a weight, and fighting it becomes exhausting and painful.

REACTIONS DUE TO STRESS AND THREATS

How do we stop this? Speaking up is not easy; it is unnerving. When faced with a threat, there are generally three options: flee, freeze, or fight. You can choose to walk away and seek employment elsewhere, but does this really help anyone in the long run? By quitting, you leave the problem for someone else to possibly face. Alternatively, you can do nothing, continuing to endure degradation, unfair treatment, threats, etc., which will almost certainly cause you harm over time and be a detriment to your health. Nick's continuous undermining and threatening comments gradually weakened me. However, if you are willing and able, standing up to them is another option. It requires strength, courage, diligence, and perseverance. I had to choose several times not to back down, committing to doing whatever was necessary to hold HDR accountable for their actions. They lied to my face, manipulated me, and gaslighted me for not just months, but years. For anyone deciding to fight, the most crucial elements might be the will to know the truth and the determination to never let go of it.

Around this particular time at HDR, my response was somewhat of a combination of all three options. After the meeting with Jessica and Nick, I still felt the need to protect myself and fight for the truth, as they were clearly not presenting it. So, I forwarded Jessica the email where Nick threatened my job.

It was also around this time that I consciously decided to avoid reaching out to Nick as much as possible. I resolved not to email, IM, text, or call him unless absolutely necessary. I even changed his contact name in my phone to "Nick Nooooooo Stone" as a reminder to think twice before contacting him. I couldn't endure his treatment any longer. Instead, I worked through tasks more independently, knowing that although I could likely be more efficient by asking a question, it was better to navigate tasks alone than to face harassment. This was my way of avoiding or fleeing from him.

At the same time, I wasn't leaving the company. I believed in HDR and was not willing to leave because of this situation. Although it was not quite the same as the times I froze during instances of sexual harassment from him, this strategy could still be considered a form of freezing. However, I did decide to focus on myself and my personal life, starting to exercise more and engaging in other forms of self-care.

You can make any of these choices when faced with a threat. Even a single celled organism has these three basic choices: move forward, move backwards, and don't move. In times like this, I think the best decision that I made was to take care of myself.

In addition, revisiting the personality results from my People Styles at Work assessment, I was defined as an analytical driver. However, I also mentioned that I consider myself more of a driving analytical, and there's a reason for this distinction. After receiving my results, I participated in a four-session training course at HDR. One of the sessions explored how different personality styles react under stress, demonstrating that individuals tend to shift from one style to another in a zigzag pattern as stress increases. My observations of my own reactions to stress are why I believe I am actually a "driving analytical." Under low stress, I am inclined to analyze, learn, and grow. Introduce a slight increase in stress, such as a work deadline, and I spring into action. This is my comfort zone; give me tasks with deadlines, and I will efficiently oscillate between these two modes to deliver quality work without any issues. The third stage for a driving analytical personality shifts towards amiable traits. In this state, I am agreeable and lean towards avoiding conflict. Even under this significant stress, I strive to be nice, kind, and prioritizing others' needs over my own. However, push me further into stress, and I become expressive. This is the fourth and final level of stress, a stage that the HDR instructor emphasized caution about, essentially advising to "watch out" if you or someone else reaches this point.

It was at this fourth level of stress when I requested a new supervisor, and this high-stress state influenced some of my reactions in meetings about these complaints. HDR allowed this to happen, leaving me in this debilitating state for extended periods, indifferent to the toll it was taking on my health and wellbeing.

THE MARKETING FLYER

Even in these moments under extreme pressure, my intentions were still honorable and committed to doing the right thing. But what about the rest of them? Were they devising strategies to help me navigate the evident challenges? Or were they orchestrating more forms of retaliation?

Within a week after the conference room meeting with Jessica and Nick, Laurence made alterations to a marketing flyer about our group. He updated Marcus's title and increased the years of experience for himself, Thomas, and another colleague from a nearby office. Strikingly, he reduced my years of experience by one. This act was a direct form of retaliation and illegal.

Fortunately, I was included in the email by the employee who revised the flyer when it was sent to the office staff for review. As soon as I opened the PDF, the discrepancy in my years of experience was glaringly obvious. I promptly emailed the employee, inquiring why this change had been made, and he forwarded me Laurence's markups. Seriously!? Confronting Laurence about it, he claimed it was a mistake and offered an excuse. However, it was clear that this was a deliberate act of malicious retaliation. The animosity of Nick, Laurence, and Mitch towards me was already fueling retaliatory actions, but now they were escalating their efforts. However, the extreme lengths they were going to added more evidence to the ongoing list of what was truly transpiring.

FOLLOWING UP ON SENIOR MECHANICAL

The overarching message that I hope comes through is my firm belief that everyone has the right to decide their future and career, regardless of their gender, skin color, age, background, or any other characteristic. If you are committed, capable, and put in the necessary work, nothing should hold you back. It was during the spring of 2022 that I revisited the topic of the senior mechanical role, a position I had applied for over six months prior without receiving any updates since. I initiated by contacting Nick and Mitch, referencing the role and expressing that it remained a goal of mine. I requested information on the next steps required to attain this position and urged them to clearly outline these steps. Something that Jessica had recommended.

There is a certain positivity that churns within me. A belief that with persistence anything is possible. My beliefs are not just confidence in myself, but even confidence in the abilities of others around me.

I must acknowledge that Jessica did offer some support along the way. She encouraged me to pursue this inquiry for the promotion requirements, assuring me that I had the right to do so and that they would be obliged to respond. Unsurprisingly, Nick did not provide any such roadmap. I persisted, following up every few months.

As spring transitioned into summer, I was officially assigned Noreen, the office manager, as my mentor. Given the ongoing circumstances, I wasn't sure what to make of this development. It had been about a year and a half since my initial complaint about Nick. I chose to trust Noreen and approach the mentoring relationship with a goal of maximizing its benefits.

I expressed to Noreen my overall satisfaction with HDR as a company. I conveyed my intention to remain with HDR for the long term and my interest to advance into the senior role within our movable bridge group. I also inquired about her career journey, the challenges she faced along the way, and how she navigated being treated differently as a woman. Noreen shared her perspective, noting that she believed each generation was making strides toward greater equality.

I also took the opportunity to try to clarify with Noreen how her role as my mentor would be distinguished from her responsibilities as office manager. Noreen stated that the two roles should generally not conflict with each other. However, in retrospect, I question if this arrangement was strategically set by the company.

FRAUD

While the corporation itself possesses extensive resources and power to protect the company at practically any cost, it is ultimately the management that controls the situation within their realm of influence. During my tenure at HDR, under the supervision of Nick, Mitch, and Laurence, I was forced into a state of constant vigilance once they began to conspire against me. Laurence had ostracized me, Mitch had begun manipulating situations, and Nick's demeanor, though his behavior slightly altered, was still apparent as if he was merely biting his tongue while they continued their schemes behind the scenes. The situation had escalated instead of improving.

But what exactly constitutes fraud? Fraud is defined as wrongful or criminal deception intended to result in financial or personal gain. In this case, they are not only lying; they are benefiting from their falsehoods. This is how far they went in their retaliation against me, with Laurence bearing the brunt of responsibility for these actions, as evidenced by the events that would unfold.

Bridge inspections in the U.S. are governed by the Federal Highway Administration (FHWA), a federal agency. The FHWA mandates the presence of a qualified Team Leader during any bridge inspection. To be deemed a qualified Team Leader, one must complete a two-week training course provided by the FHWA through the National Highway Institute (NHI), in addition to fulfilling a series of other requirements.

Within our movable bridge group, which had secured numerous bridge inspection contracts, I was the individual who was designated with these qualifications, thus enabling us to procure these contracts. I was the only qualified Team Leader actively participating and engaged in these inspections for HDR.

Every time Laurence and Nick barred me from participating in these bridge inspections, they were engaging in gross mismanagement and malfeasance. However, the situation escalated to fraud when Laurence responded to our client via email. For fraud to be considered a criminal act, there must be evidence that the perpetrator knowingly misrepresented themselves or the truth, which is what Laurence did. He utilized my credentials to secure contracts while refusing to allow me to perform the work.

By the summer of 2022, another bridge inspection was imminent, scheduled to take place in less than a month. The client, in this case the Florida Department of Transportation (FDOT), reached out to request a list of Team Leaders for the contract. As we were a subconsultant on this project, the prime consultant forwarded the request from the FDOT. In response to the prime consultant's specific inquiry, "Does anyone have the 2-week NHI course to qualify them to be team lead?" Laurence undeniably responded, "Diana is the only one."

It had already been months since I had been removed from these inspections; at least eight inspections had transpired in my absence. There was even a possibility that someone else, potentially the prime consultant, had reported the situation, prompting the FDOT to request the list of Team Leaders. Although I cannot confirm this, what I am certain of is that Laurence and Nick proceeded with the subsequent inspection without a Team Leader present on site, violating Local, State, and Federal government regulations.

MORE HARASSMENT, COMPLAINING, AND RETALIATION

Let's take a step back to recount an incident from two years prior, in the summer of 2020, which I have yet to disclose. Nick and I had traveled to a location near the Kentucky and Indiana border for a bridge rehabilitation inspection. Upon arriving at the hotel, Nick instructed me to visit his room. Assuming it was work-related, I complied. However, upon entering, he merely made a show of spreading his arms across the bed, patting it suggestively as an invitation to sit. No specific reference to work was made and the atmosphere became uncomfortable. His demeanor was disturbing, and his gaze lingered inappropriately. Feeling uneasy, I excused myself, citing the need to prepare for dinner.

Later, after dinner, I texted Nick to confirm the meeting time for the following day. He disregarded my question, responding with a curt "come over" instead. It was nearly 10:30pm; what could he possibly want at this hour? Unsettled, I replied, "I'm so tired," attempting to deflect his invitation. His response, "So no? Door is already open. Now I have to get up and close it," was loaded with expectation and disappointment. His behavior, beginning even hours earlier, was unprofessional and sexually charged.

The following day, Nick's aggression was palpable. While at the airport awaiting our return flight and desperate for a neutral topic of conversation, I brought up the Women's Employee Network Group that HDR was in the process of launching. At this stage, the initiative was still in its infancy, with internal emails circulating about its formation. This was the same group that Noreen later recommended I join. However, when I mentioned it to Nick, he responded with a veiled threat, implying that if my involvement in the group detracted from my responsibilities within our team, I would find myself out of a job. His words were another threat to my employment and an attempt to intimidate and control me.

This was a response that Nick often had. He would make a sexual advance, and when I rejected or ignored him, he would feel scorned and retaliate. This seems to be a typical occurrence, especially in cases of sexual harassment. In addition to it being retaliatory, it is an objectification. As if there was some ownership over me, and if I didn't play along I'd get tossed to the side like a toy he no longer wants to play with.

Hugs from Nick were also a source of angst and sexual

harassment as the embraces were creepy, prolonged, and overly handsy. He would caress my back and arms, causing me to need to pull away and recoil in discomfort.

Fast-forwarding to now the summer of 2022, I had since lost the weight that I had previously gained, and Nick resumed his sexual insinuations. We bumped into each other in the hallway at the office, and he invited me to join him in one of the small conference rooms nearby. It was here that he finally informed me that he and Mitch had decided to promote me to senior mechanical, though the decision still needed to be processed by Noreen. He assured me, however, that I had met all the requirements for the position. Then, he gave me one of those unsettling looks. I could tell from his body language that he was expecting a hug. This time, however, I maintained my distance, extending my arm out indicating I did not want a hug, and simply thanked him verbally, expressing my gratitude about the promotion.

Within the next few days, his aggression escalated to an extreme level, prompting me to reach out to Jessica for support. We discussed a specific instance of his unreasonable behavior, related to a task that was nearly complete but had not yet been finalized by Nick, as required by company policy and industry standards. The issue revolved around the review process; Nick needed to complete the final step in reviewing some work we had done. As far as I knew, he had not performed this last step, so I emailed him about it. Weeks later, he responded, asking me to check whether he had completed it. The email exchange became long and frustrating as Nick either could not understand or pretended not to understand that there was no way for me to know whether he had completed the final step. It would be comparable to asking someone, "Can you tell me if I locked my door this morning before I left my house?" How could I possibly know? This is another tactic of retaliation that managers might employ, assigning "impossible tasks" to report poor performance, setting you up for failure.

At the time, I was unaware of Nick's motives; I merely attempted to clarify to Jessica that checking on his behalf was impossible. She steadily told me that I needed to articulate my responses more carefully and pay closer attention to Nick's writing. Of course, the blame landed on me once again. This time, however, I decided to challenge her perspective. I questioned why she was solely concentrating on my need to decipher Nick's writing, neglecting to address his failure to comprehend my messages. I conceded that perhaps I could have phrased my responses slightly differently, but the core message would remain unchanged. I argued that Nick, in fact, was the one who needed to improve his reading comprehension, as he was the source of the complication.

We also discussed the promotion Nick had mentioned, and his sudden shift back to aggressive behavior. I inquired if there were any possible reasons or justifications they might use to retract the promotion offer. Jessica explicitly assured me that my relationship with Nick would not influence the decision. She explained that only the parameters outlined in the job code description could potentially hinder a promotion. This information provided some reassurance.

Nonetheless, over the following weeks, the situation did not improve. I had reviewed another task and sent it back to

Nick, who then requested a phone call to discuss it further. During the call, he adopted an extremely aggressive tone and talked down to me. He instructed me to make changes to the review, but when I attempted to articulate my viewpoint, he wouldn't allow me to speak. His oppressive behavior and domineering mentality were evident, as he displayed complete disregard for my understanding of the situation. When I tried to persist, he threatened to strip me of my responsibilities. Feeling cornered, I backed down.

With the resurgence of aggression directed towards me and other antics, I grew increasingly apprehensive that they might attempt to revoke my promotion. To safeguard myself, I decided to put everything in writing and meticulously document the entire process. I followed up with Nick via email, inquiring whether he had forwarded the matter to Noreen and requesting an update on the situation. Additionally, I expressed my appreciation of their approving the promotion. Nick replied, confirming that he had reached out to Noreen but had not yet received a response. Two days later, during my scheduled mentor meeting with Noreen, she reassured me that she was actively working on my promotion. Hearing this was a huge relief.

It is virtually impossible to fully comprehend the gravity of such situations until you find yourself in the midst of it. I thought things were already exceedingly bad at that point, and indeed they were, but I was utterly unprepared for the severe escalation that was about to unfold. My decision to blow the whistle would set off a chain of events that, at the time, I couldn't have imagined as a realm of possibility.

6

WHISTLEBLOWING
AND MORE RETALIATION

WHISTLEBLOWING

Whistleblowers are individuals who witness unlawful activity and courageously choose to speak out about it. The significance of whistleblowing lies in its ethical nature—it represents the average person taking a stand and declaring that something is not right, legal, or safe. Whistleblowing can pertain to both minor and major issues.

However, if the complaint pertains to a major issue, such as fraud or other illegal activities, you may be afforded protection under whistleblower laws. Whistleblowing is commendable, provided it is not done in bad faith. Nonetheless, the law might offer protection against potential retaliation, which unfortunately is a common occurrence following such complaints. The applicability of these protections depends on the nature of your complaint, and on the various whistleblower protection laws available.

Whistleblowers are individuals of integrity and high moral standards. Employees who are prepared to put themselves on the line because of their principles, regardless of potential consequences. For far too long, there has been a prevalent fear of speaking out against the corruption of those in positions of power. Speaking truth to power can certainly be dangerous but it shouldn't be. Our society is in dire need of more individuals willing to take a stand to help eradicate corruption. After the fear of retaliation, the second most significant deterrent for potential whistleblowers is the belief that their actions will not lead to change. It is crucial to understand that everyone has the power to effect change, but this requires a willingness to act.

As technology continues to advance, it is becoming increasingly apparent that corruption has spiraled out of control. Unlawful and unethical acts are becoming harder to conceal. A shift towards remote work led employers to intensify employee monitoring. Whether it's through computers or phones, if the equipment belongs to the company, they likely have access to it and are monitoring it. This includes emails, instant messaging, files on servers, virtual meetings, advancements in AI, and more.

There is now more tracking data and evidence available they can collect than ever before, even if it is happening illegally with your personal privacy invaded, and even to a point that they may know exactly what is on your mind – more on that later. You can choose to fear this surveillance or recognize that it can also work to your advantage. The more whistleblowers come forward, the more we can contribute to a society and a future that upholds justice. By working to pre-vent retaliation altogether, we are creating a safer and more equitable workplace for everyone.

GOOD FAITH

We've discussed bad faith and its associations with deceit and malice. Now, let's delve into good faith. This term gen-erally denotes honest dealings, implying that your intentions and actions stem from genuine beliefs and motivations. It means striving to be fair and impartial in your interactions. Good faith is essential for fostering a trusting team. If bad faith prevails and is allowed to escalate without accountabil-ity, relationships among team members will become strained, ultimately detrimentally affecting the team.

In the context of employment, you can think of the relationship between you and your employer as a contract. They have hired you to perform specific work and complete assigned tasks, and in return, they are obligated to provide a safe working environment and compensate you in accordance with the agreed-upon terms. When broken down to its basics, this encapsulates what the relationship between an employer and an employee should entail.

Good faith should emanate from both parties in this rela-tionship. As an employee, you are expected to perform your tasks ethically and promptly. If you are investing the necessary effort, you are fulfilling your part of the contract. Conversely, as an employer, neglecting and ignoring complaints does not fulfill the obligation to provide a safe working environment. Turning a blind eye is dodging responsibility, and acting in bad faith can leave employees feeling unheard, defeated, or even traumatized if the situation spirals to extremes.

BRIDGE INSPECTION REGULATIONS

Bridge inspection provided a welcome change of pace, offering an opportunity to leave the office and work in the field—an aspect of my job that I enjoyed. As a qualified Team Leader, you are the person-in-charge holding the authority and safety responsibility during the inspections, making your presence on-site mandatory for all bridge inspections.

To qualify as a Team Leader, you must complete the NHI training course, Safety Inspection of In-Service Bridges. This comprehensive bridge inspection training course covers the necessary aspects of bridge inspections, enabling inspectors to relate conditions observed on the bridge with established criteria. While the course does not go in-depth into specialized or complex bridge inspections, such as for movable bridges, it does address the duties and responsibilities of inspectors. Additionally, the course defines concepts including personal and public safety issues associated with bridge inspections.

By refusing to allow me to participate in these inspections, Laurence and Nick were breaching Local, State, and Federal regulations that mandate the presence of a Team Leader on site. It became evident that Laurence was aware of this violation, as demonstrated by his response to our client in an email. Nevertheless, they proceeded to conduct the subsequent inspection without a qualified Team Leader present.

On the same day that I met with Noreen, who confirmed she was working on my promotion to senior level, we received an email from the prime consultant requesting information for the next inspection. This prime consultant was responsible for conducting the structural components of the inspections, while our team handled the mechanical and electrical aspects, albeit on different days. They informed us that they had completed the structural portions of the inspection and were inquiring on scheduling the date for our inspection. As they typically did, they addressed the email to both Laurence and myself, leaving Nick out of the correspondence.

In response, Laurence provided them with the timeframe for the inspection and promised to follow up with the exact date. Within two minutes, I sent a direct email to Laurence, in the same thread, inquiring whether I or Nick would be participating in this inspection. He did not reply.

BLOW THAT WHISTLE

About a week later, Nick invited me to a celebratory lunch in honor of my years of service. Just a short time before this invitation, Duane had stopped by my desk for a chat. He shared that he had received a gift in recognition of his years of service—a gesture that HDR and its managers were supposed to extend to all employees. Over the years, I had seen Randall,

Marcus, Thomas, and Duane receive such tokens of appreciation, typically a small plaque, although Duane was showing off a fancy pen he had received. Despite my three-plus years of service, I had never received any gifts. I mentioned to Duane that such gestures were not particularly important to me, yet the disparity was notable.

On the other hand, I do value being treated fairly. I couldn't help but wonder why the rest of the group received gifts while I was overlooked. The conversation must have reached Nick because he explained that the lunch invitation was his way of compensating for the absence of a gift. During the lunch, I brought up the upcoming inspection and mentioned that Laurence hadn't responded to my inquiry. I asked Nick if he had discussed it with Laurence, and he confirmed that he had. According to Nick, he would be the one participating in the inspections. When I asked why, he simply said, "Laurence wants me to go," accompanied by a smirk. At the time, I interpreted this as perhaps an ego-driven response but, in hindsight, I believe it was one of the manipulation tactics, to deflect blame away from Nick.

As time went on, Mitch and Laurence began to be used as scapegoats for Nick. Even if the blame was shifted, onto Laurence in this situation, it still lessened the amount of blame placed on Nick, potentially helping the company avoid liability. They were aware that Nick was the problem, yet they continued to assist and even empower him. They were conscious of the discrimination at play, and Nick had seemingly been advised to shift the responsibility onto others. After Nick's response, I couldn't comprehend how they would permit this, especially since Laurence had informed the client that I was the only qualified Team Leader.

Although I was frustrated, I was also too apprehensive to confront him directly about it. Instead of addressing that issue, I decided to follow up about the senior role once more, hoping to find a semblance of fairness amidst the chaos. Nick mentioned that he needed to check in with Noreen.

This was one of those pivotal moments in life, a choice that would end up irrevocably altering my future. On one hand, it was blatantly obvious that they were violating bridge inspection regulations and deceiving the client. On the other hand, my promotion was tantalizingly close, just weeks away. However, at that moment, my focus was not on the promotion; it was on the issue at hand—the inspection qualifications. They were not doing the right thing. It was fraud.

So, I took a stand and blew that whistle. First, I sent an email to Laurence, attaching the previous email chain in which he stated that I was the only one who had completed the two-week training course. I demanded an explanation as to why I was being refused to participate in the inspection. Laurence responded with a pretext, claiming he was seizing business opportunities in different regions of Florida. In essence, he was misappropriating funds from a bridge inspection contract to support marketing and business procurement endeavors. This justification does not validate non-compliance with bridge inspection regulations, nor does it constitute an appropriate use of our contractual budget.

He also mentioned that once we onboarded a new electrical engineer with a PE license, I would accompany them to inspections. This was because our existing electrical team within Florida, lacking PE licenses, was deemed ineligible for these inspections. Essentially, Laurence was unwilling to conduct inspections alongside me. This hadn't been an issue before, but it changed following my complaints about Nick. It was more retaliation.

I responded, highlighting that the Team Leader qualification was a mandatory requirement, and pointed out that the job opening for an electrical PE had been posted for at least two years with no results. Laurence countered, expressing doubt that this was a firm requirement, and stated that he would have Mitch investigate. How could it be unclear, the FDOT had recently requested these qualifications. Laurence also noted that the referenced email chain was internal communication, which it clearly was not. These email exchanges occurred on a Friday. By the following Tuesday, I followed up on the email thread, this time addressing Mitch and including the project manager (PM) on the job.

The PM intervened, clarifying to everyone involved, including Mitch and Laurence, that this FDOT District local office required the presence of a Team Leader during these inspections, as noted in the signed and executed Contract Agreement. He acknowledged that this Team Leader could be from either HDR or the prime consultant—a correct assertion. However, he mistakenly noted that this requirement applied only to the specific FDOT District in question, not elsewhere. This was incorrect; while other FDOT districts might not have explicitly stated the need for a Team Leader in their inspection contracts, it remains a requirement at both the State and

Federal levels. It was around this point that Laurence finally relented, suggesting that I coordinate with an out-of-state electrical engineer named Grant Obrien.

During this time, two email threads were active concurrently. The previous night, Nick had emailed me requesting the past inspection report, a document we traditionally bring with us to reference and annotate during inspections. I did not immediately reply to Nick's email, choosing first to address Mitch in the other thread. When I did respond, I directed my query to Laurence, inquiring whether he planned to reschedule the inspection due to an impending hurricane. Included in this correspondence were Nick, Laurence, and myself. Nick anticipated that the inspection would likely be postponed, while Laurence, who was already on Florida's east coast, contemplated conducting the inspection solo. This was a terrible idea, not only because of the storm, but also due to safety reasons, inspections should not be performed solely. I assumed this would be negated by the storm, and when he requested the electrical section of the report, I promptly sent it. Nick, subsequently, requested the mechanical sections for the second time.

This was all transpiring simultaneously with the other ongoing email chain. I relayed this to Nick, explaining that I was awaiting confirmation regarding my required presence as a qualified Team Leader and highlighted that the FDOT had recently inquired about this, indicating the necessity that I perform the mechanical inspection. I attached the previous email chain originating from the FDOT, in which Laurence acknowledged that I was the sole team member to have completed the requisite two-week course, for Nick's reference. I assured Nick that I would update him as soon as I received a response.

Nick's reply disregarded the issue at hand, fixating instead on his initial request for the report and accusing me of insubordination. This was not a matter of personal opinion about who should conduct the inspection; it was a matter of adhering to Federal guidelines as well as to the signed Contract Agreement. Nonetheless, Nick's accusation of insubordination filled me with fear.

This scenario illustrates a common plight of whistleblowers: challenging corruption when those involved hold power. They have the authority to dismiss you, label you insubordinate, and manipulate the situation to their advantage. Their corrupt practices may be so ingrained that they are even blind to the wrongdoing and malice in their actions. When one dares to blow the whistle, these individuals may react with hostility and leverage their power to retaliate.

When Nick sent his last response, he CC'd Mitch. In my subsequent reply, I included both Mitch and the PM, expressing, "I am acting in the best interest of the client and the contract, and I gave you a valid explanation. Attached is the report, per your request, however, I may be going on the inspection instead of you. I am waiting to hear back."

Mitch chimed in, sharing his interpretation of the regulations and his preferences for how we should conduct the inspections, which, once again, contradicted Federal regulations. Eventually, the PM responded to the other email thread, asserting the requirement of having a Team Leader present during the inspections.

Nevertheless, the damage was done. Anxiety induced, fueled by Nick's reactions. Employers have a way of manipulating situations in their favor, often making the employee—the 'little guy'—appear to be in the wrong. For a while, their attempts to shift the blame led me to question my actions as if I had done something wrong. Did I make a mistake? Should I have just kept quiet and thoughtlessly did as I was demanded? It was much later that I realize there's no reason to feel guilty. On the contrary, I should take pride in my actions. I stood up to them, fulfilling my duty and responsibility as the trained and qualified Team Leader to voice my concerns. This is the essence of whistleblowing—acting in good faith to expose wrongdoing.

REPORTING IT

The next step in whistleblowing is reporting the issue, in seek of correcting the problem and getting accountability as required. Properly reporting the misconduct to the correct individuals or officials is essential when whistleblowing, and is even included within and mandated by the verbiage of whistleblowing laws. This can be done internally within the company or externally to a governing agency, and it can be done anonymously or not.

I chose to start by reporting the issue internally. About three weeks after the whirlwind of email exchanges, I had my next mentor meeting with Noreen. My stress levels were still high, and I was anxious about what Nick, Laurence, and Mitch might do next.

We met for lunch, as had become customary, since it suited Noreen's schedule best. I found myself unable to eat much this day, eventually packing my leftovers to take home. During our meeting, I brought up the senior mechanical role again having at this point become focused on the role, as if the promotion would indicate that HDR was not against me and provide a sense of job security and future with the company.

Noreen, however, finally confirmed—the promotion had been approved and budgeted for. Incredibly relieved on one hand, I then broached the topic of potentially being moved to a new supervisor, suggesting that it could be greatly beneficial for me. While she acknowledged the possibility of a change in supervision, she also managed to make it seem unlikely. My head began to spin, and, overwhelmed, I leaned forward, rested my elbows on the table, and began rubbing my temples.

Noreen also mentioned that our mentoring relationship was nearing its end. She suggested that I consider signing up as a mentor instead of a mentee for the next iteration of the program. I acknowledged the merit in her suggestion, indicating that I would give it some thought.

Upon returning to my desk after lunch, I found it impossible to concentrate. My mind was consumed by the tumultuous events of the past three years and them leading to the past few weeks, including the whistleblowing. Having Noreen as my mentor, especially given her position as office manager, had begun to feel like the sole possibility of escape from this turmoil. And now, our mentoring relationship was about to conclude?

Compelled by desperation, I found myself back in her office, much like I had done a little over a year prior, seeking avenues for career growth. However, this time around, my demeanor was far from strong and confident; I was worn out, scared, and vulnerable. As soon as I began speaking, my emotions overtook me. I couldn't hold back any longer. I shared with her my profound fear of Nick and my inability to work under him any longer. I disclosed the ongoing issues regarding bridge inspection qualifications, highlighting how they were falsifying information to the client, utilizing my credentials while simultaneously preventing me from participating in the inspections.

Noreen's response was one of fury, though not directed at me. She exclaimed, "There is a disruption in the group!", declaring that she received my dire message. Nevertheless, I reiterated my fears, emphasizing that it wasn't just Nick; Mitch and Laurence were also implicated. Lodging complaints about harassment and discrimination is challenging enough, but raising allegations of fraud involving multiple individuals in upper management was an entirely different ordeal. Noreen, in a tone filled with genuine concern, said something that one should allow more than a moment to let sink in: "But you cannot continue like this." She acknowledged the severity of the situation, recognizing that it had escalated to a point where it was causing me harm and that something needed to change. I told her I would try to keep a low profile while she assured that she would handle Nick and the situation.

I had hit a breaking point. The ongoing strain had become unbearable and change was imperative. After leaving her office, the overwhelming emotions rendered me unable to remain at work for the rest of the day. I made my way to my car, and once again, broke down. I began to cry, and my jaw started chattering uncontrollably—a visceral response to the extreme distress I was under.

It was grief. Like mourning the loss of a loved one, I also began to mourn the loss of my job at HDR and possibly my career in movable bridges, which I greatly valued. Was this the end of my career? Subconsciously, I believe I saw it as a possibility although not something I was willing to face. At the time, I wasn't fully aware of what was transpiring or how long it would last.

MORE STRESS AND COMPLAINTS

During this very week, Nick, Mitch, Laurence, and Duane attended the 2022 Biennial Movable Bridge Symposium, organized by a non-profit organization called Heavy Movable Structures (HMS). I had initially inquired about attending the symposium due to its wealth of educational presentations and the attendance of vendors, manufacturers, contractors, and other consultants from our industry. Duane, who was one of the presenters, had expressed gratitude as I helped him prepare. I was happy help and wished I could have seen his presentation.

I first approached Laurence about attending the sym-posium roughly a year in advance, to which he responded that we would have to wait and see. Six months prior to the event, I asked Mitch, who informed me that the group had a limited budget for the symposium, effectively ruling out my attendance. Interestingly, when I was in Noreen's office, she appeared perplexed as to why I was not at the symposium with the rest of the team. A reaction I find strange in retrospect, considering she would have been in charge of the budget.

Amidst the overwhelming stress from all these events, and contrary to my intention to keep a low profile, I emailed Nick and vented my frustrations. However, I did so in a way that was "ask assertive." I peppered him with numerous questions regarding a specific task he had assigned me. This happened while Nick and the others were at the HMS symposium. Despite this, Nick responded to my email and CC'd the PM for this particular task. This was someone whom I had not worked with before, marking our first interaction. Nick was setting the stage for a poor first impression and attempting to disparage me. It was unnecessary for him to include the PM in this exchange. The email chain continued back and forth for a bit. Reflecting on this period, I should have refrained from contacting Nick at all and kept to my own intentions of laying low.

Just a few weeks prior to this, around the time I was processing the data and working on creating the report for the last inspection Nick had performed before I blew the whistle, I had a concerning dream or nightmare. This inspection was of a bascule style bridge which as I have mentioned include large concrete counterweights at the one end to create an efficient system while the bridge operates, rotating like a seesaw. Sometimes these counterweights are incorporated above the deck in plain sight and other times are below the deck, hidden from view. In the case of the latter, there is also often construction of what's called the counterweight pit within the substructure bascule pier. When the bridge is in the open position, the counterweight ends up tucked into this counter-weight pit, an area you never want to be in during operation as you could get killed. The dream consisted of Nick and I on an inspection of this type of bascule bridge. In the dream, Nick directed me to head down into the counterweight pit to look at something. Following his orders, I climbed down and the next thing I knew, the bridge started to open. As I found myself trapped, I looked up and saw him starring down, watching me from above as I was about to get squashed by the counterweight. This was phycological, it was a nightmare, and I woke up feeling scared.

In the midst of the current chaos, I reached out to Jessica once again. I reiterated how extremely stressful it was work-ing with Nick and requested a change in supervisors for the second time. I informed her that I had also discussed this mat-ter with Noreen. I shared that Noreen had acknowledged the current setup was no longer a viable option and that I couldn't continue in this manner. Jessica assured me she would check in with Noreen regarding this issue.

I also expressed to Jessica my belief that Nick did not have good intentions when it came to me, and I conveyed, yet again, my lack of trust in him. I mentioned that even sim-ple tasks, like filling out my weekly timesheet, had become sources of stress knowing he was approving them. I recounted how, at times, I would feel a surge of anxiety just from seeing his name pop up on my phone during an incoming call.

I then clarified my commitment to my career path and my desire to continue working within the movable bridge sec-tor. When Jessica inquired about my willingness to relocate, I told her I would prefer not to at this time, questioning why I should have to upend my life because of his unjust treatment and behavior. I mentioned that I enjoyed my home and where I lived to which she responded asserting that relocation was not necessary and sharing her own experience of declining requests from HDR to move to Omaha, Nebraska, where the company's headquarters are located.

I proceeded to tell Jessica that I was open to collaborating with other offices, and I informed her that Noreen had confirmed the approval and budgeting for my transition to the senior mechanical role. I suggested that this transition, slated for the end of the year, could be synchronized with my change in supervisors for a smoother process. Jessica concurred, agreeing that it seemed like the most logical course of action.

Furthermore, I disclosed to Jessica my recent act of whistleblowing regarding the inspection qualifications and my report to Noreen. I explained how they were misusing my credentials for a contract while preventing me from participating in the actual work. I offered to provide more information if necessary. Jessica responded that it wasn't needed at the moment and reassured me that they couldn't retaliate against me. At the time, I didn't fully grasp what that meant. Nonetheless, I emphasized that it felt like they were already holding a grudge against me.

MORE RETALIATION AND I GET MOBBED

Looking back, I realize that many times over they acted with intentional malice. My complaints had been neglected for far too long, and things continued to escalate as they relentlessly pushed the boundaries, concocting ways to blame and trap me.

It had been around a month since I blew the whistle, and Florida had just endured the devastating impacts of a hurricane. The rescheduled inspections were approaching, and this time, I was set to conduct them with our out-of-state electrical engineer, Grant, instead of Laurence. I had previously worked with Grant on several inspections before Nick and Laurence removed me from these duties almost a year prior. Grant, hailing from the Midwest, had always been respectful and gentlemanly. However, the person I met this time was a far cry from the gentleman I remembered.

Before even heading out, Grant was highly non-responsive to my coordination efforts for the inspection planning. I had sent him details like the travel booking for the hotel location, but he failed to respond. I believe I sent him three separate emails. Since he had already confirmed his availability, I assumed he was just busy and trusted that he was capable of handling his own travel arrangements.

He finally responded on the morning of our first inspection. I reassured him that I was still on track for the planned meeting time and had even sent him the location of where we would park and meet. I was around twenty minutes away from this meeting location, and my GPS indicated that I would arrive right on time. That's when he called me, stating he wanted to grab some lunch and would need extra time. I agreed that it wasn't a problem and took the opportunity to grab a quick snack myself. Despite the adjusted meeting time, I arrived and parked early. To my surprise, Grant was already there. Then, his first words to me were, "You're late." I was taken aback. He was the one who had requested to adjust the meeting time, and now he was accusing me of being late?

Throughout the rest of the trip, he continued to behave in this peculiar manner, as if he was meticulously cataloging items to report negatively on me. He shadowed me during the inspection, making me feel distinctly uncomfortable and even brought me to tears. It was clear that he was not acting like his usual self. Since Grant had also attended the HMS symposium a few weeks earlier, they may have conspired against me during that event, recruiting him to participate in their retaliation.

In addition to navigating Grant's unexpected and rude behavior, I was also contending with Nick's attempts to contact me. He inquired if I could find time to review a calculation submitted by a contractor. I told him I might be able to do it from the hotel or later that week, but he decided to take care of it himself instead. On Friday, after returning from the inspection, he sent me a text asking if I had reviewed the calculations before a meeting that afternoon. I reminded him that he had told me he would do it, realizing that this was another one of his retaliatory setups.

I revisited the email chain from the previous week—the one where I had been "ask assertive" in response to the overwhelming stress. I added to it, expressing my extreme frustration working with him, accusing him of playing games and being ridiculous and unprofessional. I told him that his behavior generated a lack of trust and stated plainly that I didn't believe he had good intentions regarding me and my career. I also mentioned that I had been dealing with this for over three years and had already discussed the matter with Jessica and Noreen. Yet, despite all this, the situation persisted.

When the afternoon call came around, I decided to participate, even though I wasn't in the best frame of mind. When asked about the hydraulic system calculations and specifically about the pressure drops for the components, I apologized, explaining that I hadn't been given sufficient time to review the calculations. However, I did my best to provide an initial assessment on the spot, noting that the calculations seemed reasonable, with variations likely based on the differences between our assumed components and the ones they had selected for the system.

The following week during our weekly progress meeting, Nick brought up that Thomas was low on work. When it was my turn to speak, both Nick and Mitch inquired whether I had any work I could delegate to Thomas. Mitch hadn't consistently attended our progress calls, but he had started to join more frequently amidst the train of ongoing harassment, discrimination, and retaliation. I paused, taking a moment to think before responding to their question, my subconscious recognizing it as another potential trap. After careful consideration, I mentioned that there was a part of one of my tasks I could potentially hand over to Thomas, if necessary. Mitch then backtracked, suggesting I take more time to think it over and get back to them later.

By this time, the opposition against me wasn't just limited to Nick, Mitch, and Laurence; it also included Grant and others I had spoken to. I was facing mobbing. I tried reaching out to Noreen, but her attitude towards me had also changed. The last time we had talked was when I reported the whistleblowing incident, and she had been supportive. Now, however, she too was posing leading questions, one of which was about how much longer I saw myself working. She even attempted to put words in my mouth, suggesting it might only be a few more years. Plain gaslighting. I corrected her, stating that I saw myself working for another 30 years, and emphasizing my desire to continue my career with HDR. I wanted to make it clear that I liked the company and had no intention of leaving.

Some individuals are more experienced with their attempts at gaslighting, making it harder to perceive, like with Jessica, but others will just blatantly throw out a direct lie and hope it sticks. That was what Noreen was doing. It was around this time Noreen also attempted to gaslight me with statements that I wanted to work in different industries and even that I had been in a severe car accident and had suffered a head injury, none of which is true. It was an indication of their knowing what effects could arise from this situation.

I REPEAT: HR IS *NOT* YOUR FRIEND

By the end of the week, I found myself back in Jessica's office. We discussed the email chain where I had followed up highlighting Nick's lack of good faith towards me and my lack of trust in him. Nick had actually already forwarded the entire email chain to her prior to our meeting. I expressed to Jessica how I had been communicating these issues to them for a while and emphasized the immense stress it had caused me. I also delved deeper into the whistleblowing incident related to the inspection regulations, discussing the numerous email exchanges that occurred in its wake. She attempted to downplay the situation, stating that those inspection qualifications weren't a big deal, but she was gaslighting, her attempt to minimize the gravity of the illegal and fraudulent activities. I couldn't help but get emotional as I told her again how this was affecting my health, including my ability to sleep.

Jessica then suggested that I write an apology email, subtly shifting the blame onto me. She told me if I didn't apologize it would be bad for my career. Being completely manipulated, I agreed, though I wasn't in a good state of mind when I wrote those emails. I decided to take the weekend to think about it. When she asked me if I still had respect for Nick, I was honest and admitted that, by this point, I didn't. I had previously respected him, but that respect had diminished over time due to his prolonged mistreatment.

Very suddenly, Jessica's demeanor changed, and she became visibly angry, raising her voice, her mouth wild open and head weaving as she insisted that I take the weekend to carefully think this over. I cringed in response. It was an obvious attempt at intimidating me while I was clearly distraught, further contributing to the mobbing I was experiencing. Completely taken back, I inquired about the senior mechanical role and the promised promotion at the end of the year, only to be told that it was no longer happening. This was in stark contrast to her assurance just a few weeks earlier that my relationship with Nick would not impact the promotion in any way.

Feeling cornered, I told her that I felt I was being bullied. At the time, I still didn't understand my rights and the law and the gaslighting they had been subjecting me to along the way only made this misunderstanding worse. When she asked for examples of the behaviors I was referring to, I mentioned a lack of empathy, lack of support, and incivility, watching as she took notes in her notebook. But what was actually happening is severe retaliation and mobbing, resulting from the whistleblowing. Despite her intimidating demeanor and manipulative tactics, I left her office and later forwarded her the email chains documenting the whistleblowing incident, as well as Laurence's response to the client confirming that I was the only one with the necessary two-week course certification. These types of documenting actions are critical in holding employers accountable and ensuring there is a record of what has transpired.

When I got home that day, I called Mitch, still distressed, and told him repeatedly, "I don't feel safe. I don't feel safe. I don't feel safe." I couldn't discern whether he took me seriously, but I knew I couldn't let them continue to treat me this way. Something had to change. In Noreen's words, I could not continue like this.

The following week, I had another follow-up meeting with Jessica in her office. Looking back, I realize that these meetings were both in her office, leaving the trail of our discussions only to my notes, further emphasizing the importance of documenting. We revisited the email chain and her forceful suggestion that I apologize. I told her that I planned to send an apology email but would specify that, in retrospect, it was an overreaction in response to that specific email chain. She implied that sending the apology was a good idea. In reality, I should not have been the one apologizing. It was just another way to shift the blame onto me, rather than acknowledging the reprehensible treatment I had been subjected to. This apology was later used against me.

Additionally, I shared with Jessica some examples of how Nick treated Thomas very differently than he treated me. I mentioned that Nick is much more respectful with Thomas and willing to review his work in a timely manner, where I had on multiple occasions been forced to wait several months for Nick's reviews. I also noted the recent group call where Nick and Mitch asked me to delegate some of my work to Thomas, describing how the way they went about it felt like a trap. I further noted that the behavior seemed to be "spreading" like a virus, affecting Mitch and Laurence as well, as I had begun to acknowledge their unfair treatment towards me too.

I've said this before and I will say it again, sometimes it's hard to see what's happening while it's occurring. Mitch and Laurence had joined Nick in the retaliation against me way before this conversation, but it was difficult to acknowledge

and admit it right away. It's a frightening thing to do, especially when the individuals involved are your management
team acting in misconduct.

MORE UNLAWFUL AND UNETHICAL CONDUCT

The very next day after my talk with Jessica, we received an
email from the prime consultant we were working with on
another FDOT District project for inspections. This was the
district that the HDR inspection PM had incorrectly stated
did not require a qualified Team Leader. This information
was inaccurate; despite what the contract stated, Federal and
State regulations mandate that a Team Leader be present
on-site during bridge inspections.

The email was addressed to both myself and Laurence,
and they were seeking confirmation of the inspection dates.
I wrote to Laurence internally, also CC'ing the PM and Nick
in the email, asking him if he had any scheduling conflicts
and stating that my schedule was open. He responded, saying
he planned to conduct the inspections with Nick and would
respond to the prime consultant accordingly.

At this point, we were once again facing gross negligence
and mismanagement. I was beside myself but decided to hold
off before responding again. It was November, and these
inspections were scheduled for January, so there was still time.

WHEN WILL IT END?

Despite everything that had already transpired, all the complaints, and all the reporting to HR, Nick's discrimination still continued. If this isn't a hostile work environment, I don't know what is. We had a railroad client for whom we consistently performed rehabilitation designs, and this particular project required almost weekly virtual calls for internal coordination. We were modeling this design using 3D software, and our movable bridge group hadn't undertaken a project of this scale before. Around this time, during two separate calls, Nick dismissed my contributions.

On one call, we were discussing an issue reported by the contractor from the field. This was a swing span bridge, and they were encountering problems with the machinery at the end of the span. The ends of a swing span on a railroad bridge have special joints for the rails that overlap, which are lifted or pulled during operation to eliminate interference while the bridge swings open. There are also wedges at the ends of the span to provide bearing with positive reactions for transferring the load from the superstructure to the substructure, while also preventing uplift. These wedges are also pulled and driven when opening and closing the bridge, respectively.

The contractor noted that it seemed the rails at the joints had been expanding unevenly, resulting in a slight curved shape due to sun exposure on one side—a phenomenon Nick referred to as "banana-ing," an exaggerated comparison to a banana's shape. Based on the problem's description, I noted during the call that this issue could cause additional unintended friction on the sides of the wedges. I expressed uncertainty about whether the design could handle the additional resistance from this friction. The end machinery was being driven by relatively small tie rod hydraulic cylinders. Our design incorporated one cylinder to drive a shared shaft for operating both the rail lifts and one of the end wedges, while the other end wedge was to be operated by its own hydraulic cylinder. My concern was that the additional friction on the side of the wedge could increase the system pressure above what the cylinders were rated for.

Nick completely dismissed me and began talking over me. He then proceeded to generally reiterate what I had just said, but did not seem concerned about it. These calculations were part of a set that I had sent him nearly two months prior, which he had yet to review. By the end of the call, still concerned about the additional friction—having been the one who performed the design calculations—and upset over Nick's demeaning and degrading dismissal, my voice raised. I made a statement along the lines of, "You can't just use those hydraulic cylinders without checking for the additional friction resistance first!" Once again, I was transparently dismissed.

On the second call, I was still upset about the previous call, and still not myself since blowing the whistle and everything else that was ongoing. I tried to discuss the same topic of friction and the calculations for checking the cylinder, unsurprisingly Nick wouldn't allow it. Given that this was a coordination call, we—meaning the structural, mechanical, and electrical engineers—were supposed to be collaborating. He allowed others to speak, but when it came to me, he insisted that we would speak separately.

Feeling rundown, I called Mitch for information. I asked him what was going on with the senior mechanical role. Mitch told me that he had sent the email chain, where I had expressed my lack of trust in Nick and followed up with an apology for my reaction based on Jessica's influence, to the HDR Business Class Group. After he sent them the email chain, they subsequently denied the promotion—a promotion that had already been approved and budgeted for. This was outright manipulation and retaliation with adverse action denying me the promotion.

Mitch then mentioned that he would be coming up to Tampa from Fort Lauderdale for my upcoming annual review the next week. Still naïve, I asked if it was a "stay conversation"—an HDR term referring to conversations supervisors were supposed to routinely have with their employees to increase employee retention, a topic covered in HDR training courses I had taken. Mitch didn't even answer the question, leaving me to get off the phone thinking they might try to fire me. I was so scared I decided to email Jessica, telling her I wanted to go over a list of items that had happened over the last two years. I was protecting myself.

While whistleblowing is typically a moral choice made to protect others or serve the greater good, with the public in mind, when you are being discriminated against, harassed, or retaliated against, it becomes more personal. It turns into self-preservation because the victim is being abused. In the meantime, I still continue to work and deliver on my tasks even though I am not feeling well.

A few hours later, I had another call scheduled. It was again for our railroad client but this time with the client themselves and Nick not included in the call, thankfully. The call went really well. I had an opportunity to present my work to the client while screen sharing. They were happy with my work and progress. The client directly told me on the call that he knew how hard I always worked. It felt so nice to hear after having to constantly deal with Nick. It also felt so good to be on a call and not have to worry about Nick putting me down and treating me poorly in front of others. It felt liberating.

The next day I emailed Jessica the list I had mentioned. In one of our last meetings in her office, she had asked me to clarify what I meant by Nick's behavior. This list should help clarify, as it was of several incidences of Nick's behavior that I had been documenting for a while. This list in writing may have prevented them, at almost the last minute, from what seems were their true intentions of firing me. But instead, it is what triggered the next phase of this whole ordeal: the internal investigation.

7

INTERNAL INVESTIGATION, ADVERSE ACTION, AND FINDING THE LAW

CORPORATIONS WILL MANIPULATE THE LAW IN THEIR FAVOR

Corporations have significant means at their disposal, while the employees are extraordinarily disadvantaged. It is like the story of David and Goliath: the employee represents David, just an average individual, and the powerful corporations are the giant Goliath. One of the ways they are advantaged is that they know the law much better than the employees. They have corporate lawyers on board and HR departments also working to protect the company, allowing for much more power over any unlawful and unjust actions that occur within the workplaces.

The lawyers and HR representatives are aware of employment laws, as well as the relevant case laws and judgments. There is case law with Supreme Court judgments from cases such as *Faragher v. City of Boca Raton* and *Burlington Industries,*

Inc. v. Ellerth, both from 1998. Corporations and employers will misconstrue and use these judgments in their favor, while the employees are completely unaware and naïve to all of it. This enables the employers to get away with bad faith actions, causing significant stress and harm to their employees.

FARAGHER V. CITY OF BOCA RATON

In this case, a woman named Beth Ann Faragher was a lifeguard for the City of Boca Raton, FL. She resigned and brought action against the City and her supervisors for the alleged sexually hostile environment that her supervisors had created. Faragher asserted that this conduct constituted discrimination in violation of Title VII of the Civil Rights Act of 1964. The defense argued that the harassment occurred outside of the scope of employment, that the complaints of harassment made could not be imputed to the City, and that the City should not be held liable for negligence in failing to prevent it.

The court's judgment held that an employer is vicariously liable under Title VII of the Civil Rights Act of 1964 for actionable discrimination caused by a supervisor. The court also held that such liability is subject to an affirmative defense, looking to the reasonableness of the employer's conduct as well as that of the plaintiff. The first judgment's importance lies in the words "vicariously liable" and "by a supervisor." This means that the employer is to be held responsible for the discriminatory actions of its supervisors. The second judgment, though, leaves the cases and liability to also be determined based on the "reasonableness" of the employer's and victim's actions, or reactions. This makes a deciding factor in liability for the conduct to also be about the responsive actions. In this case, was Faragher or the supervisors who were at fault? Or was it the City?

As a result, employers have since been working to place blame on the employees suffering from the discrimination. Instead of treating complaints for what they are and helping the individuals as required and in good faith, they evaluate what is more beneficial to their business in the long run. The employer then uses their power to manipulate situations, discussions, and evidence to suit their needs in defending possible future lawsuits, in the event that the complainant files charges.

In the case in question, the City of Boca Raton entirely failed to disseminate its policy against sexual harassment among the beach employees. This is why many employers require employees to watch sexual harassment videos during onboarding or annual training, which usually look like they were made in the 1990s. Additionally, its officials made no attempt to keep track of the conduct of supervisors. The City was negligent, being found not to have exercised rea-sonable care to prevent the supervisors' harassing conduct. In this sense, an employer now only needs to prove that they exercised a reasonable level of care when dealing with such cases involving harassment or other forms of discrimination, and that the complaining employee did not act reasonably. However, the line of what is reasonable and what is not, is not well defined and must be judged on a case-by-case basis, leaving some employers working only to make it look as though they responded reasonably while simultaneously attempting to push the victim to a breaking point.

BURLINGTON INDUSTRIES, INC. V. ELLERTH

This is a case where a woman named Kimberly B. Ellerth worked for a company called Burlington Industries for 15 months. Her employment ended with her quitting after alleging sexual harassment by her supervisor. While employed, she had been promoted once and could not provide any tangible evidence of retaliation. However, based on the judgment, it was decided that the environment had become hostile, leading to Ellerth's "constructive dismissal". The term constructive dismissal means that the employee has been forced to leave due to the unsafe and "hostile work environment".

The judgment by the court held that an employer is subject to "vicarious liability" for a supervisor that creates a "hostile work environment" for employees they have "authority" over. Again, this judgment, using the term vicarious liability, places the responsibility and liability for the hostile work environment created by the supervisor on the employer. It also dictates that the supervisor is someone who has authority over the employee, regardless of whether that authority is direct or indirect.

The court also noted that in cases where harassed employees suffer no "job-related consequences", employers may defend themselves against liability by showing that they reasonably acted to prevent and correct any harassing behavior and that the harassed employee failed to utilize their employer's protection. The term job-related consequences, is another way of saying "retaliation with adverse action", which is retaliation taken to a point that it negatively and financially changes the conditions of employment such as by demoting, failing to promote, or firing. In this case, Ellerth was promoted and later forced to quit because of the hostile work environment, or forced into constructive dismissal, which was here also considered a job-related consequence, preventing the employer from using the above defenses.

Assuming no job-related consequences, employers can attempt to establish a defense on the basis of the two necessary elements. The first is providing evidence that they acted quickly. The second is proving the plaintiff or employee unreasonably failed to take advantage of the employer's protection. Both of these elements need to be proven for this defense to be valid. Note the use of the term "quickly", again adding subjectivity as these lines might be drawn on a case-by-case basis.

FARAGHER AND *ELLERTH* CASE TAKEAWAYS

Both the *Faragher* and *Ellerth* cases have had a huge impact on current work environments within this country. In both cases, it was determined that the employer is vicariously, or indirectly, liable for the supervisors' actions. If we think of the company as an entity, these vicarious liability judgments seem to have made companies anxious about potential future lawsuits. Any sign of possible discrimination claims raises a flag. Once the flag is raised, it is very unlikely to come down. This is when these types of toxic employers may act in bad faith to protect the company by discrediting and pushing the complaining flagged employee out, disregarding the health and well-being of that employee. Because of these cases, the employer sickeningly will protect, praise, and even promote the harassing supervisor because they are explicitly vicariously liable for them.

However, the majority of employees do not want to take legal action against the company for the harassing or discriminatory conduct they are subjected to. They just want the misconduct to stop. The employee wants to be treated with respect and dignity, as they deserve to be. It is an inherent human right. Many of the cases that do end up in court are more likely filed because the employer refused to take "reasonable" action on the complaints and instead acted in bad faith and negligence. This lack of accountability and further retaliation can be even more damaging to the employee than the original misconduct. First, the employee is subjected to misconduct, then neglected, and further retaliated against by others causing additional stress and harm. It is systematic and egregious, and it is NOT the employer taking reasonable action or care to protect the employee.

Another major takeaway comes from the *Faragher* case and the failure of the City of Boca Raton to disseminate their sexual harassment policy. This brings us back to those sexual harassment training videos. Several states now require employers to train their employees on sexual harassment; however, not all states have this requirement. These trainings are usually very outdated, having been created in the 1990s. It wasn't something that I thought about until I had to, after having gone through everything I did at HDR. These videos focus heavily on sexual harassment and barely touch on other types of discrimination.

This is because they do not enforce requirements on the inclusion of all employee discrimination rights in these training videos. I was naïve and unaware of my rights as an employee and the ways that we are currently protected by law. We should all have a better understanding of this to minimize its occurrence and the harm it causes to the individuals who are subjected to it. In the same way that the City of Boca Raton did not provide a reasonable level of care with regards to their sexual harassment policy, in my opinion, HDR did not provide a reasonable level of care with regards to my training or understanding of psychological harassment, discrimination, and retaliation policies.

What Jessica did was send me a copy of the harassment policy in writing, which I did not even understand, and then schedule a conference room meeting where she worked to manipulate, intimidate, and prevent me from continuing to make complaints. It was the opposite of reasonable care; it was bad faith, retaliation, and negligence. Maybe if I had understood my rights and employment law better at the time, the majority of what I was subjected to could have been avoided.

A further significant impact made by these cases is the rule on the hostile work environment in the *Ellerth* case. It was ruled that the sexual harassment she was subjected to had created a hostile work environment that affected her ability to perform her job, forcing her to resign as a constructive dismissal. A work environment is considered hostile when harassment due to a protected class becomes severe or pervasive to the point it: affects the victim's ability to perform their job; when the environment becomes offensive, intimidating, and threatening; or when it causes a situation in which the victim's health and well-being are affected. Essentially, the reason for the *Ellerth* case term of acting "quickly", since it is affecting the employee's health. This case again makes employers vicariously liable for the behaviors of a supervisor, indirectly or directly in authority over the victim, when it creates a hostile work environment.

At the end of 2021, I requested to be taken out of Nick's direct supervision noting he was causing me harm. But Jessica immediately denied it stating that Nick was the mechanical lead in the area because she knew it didn't matter if he was my direct supervisor or if I indirectly worked with him, they could still be held liable.

Lastly, it also prevents employers from using the previous defenses when there are job-related consequences, even constructive dismissal resulting from the hostile environment. It was around the fall of 2020 into the spring of 2021 where my symptoms escalated and my health was being affected. That fall was around the time Nick arrived at the Virginia construction inspection job. This was also around when he threatened my employment via email. The way he treated me on that job, combined with his threat, left me extremely upset, prompting my first complaint to upper management. I could no longer tolerate Nick's behavior.

Despite this, I continued to work through the stress, which paradoxically drove me to work harder. However, while I remained productive, I was far from okay. Both my mental and physical well-being were suffering.

"THOROUGH" INTERNAL INVESTIGATION

At some point, your employer may decide to conduct an internal investigation, likely when you have made a complaint in such a way that they can no longer dismiss it. Mainly after a written complaint that makes it clear the behavior in question can be considered a type of discrimination. Conducting an internal investigation demonstrates that they took action once the written complaint was received. In my case, this happened when I sent a written list to Jessica detailing several instances of offensive, degrading, threatening, and fraudulent misconduct.

At the end of the day, you must protect yourself. Complaints put in writing are undeniably made; otherwise, employers may lie, deny, and twist the truth. Had I been aware, I would have submitted my complaints in writing much earlier. Once a protected complaint is made, you are protected by law from retaliation. However, it is important to note that, despite this protection, things can still escalate and become worse, even though they legally should not.

The internal investigation was conducted by Jessica, as

my HR representative. At the outset, she explained that her role was to be the "fact-finder" in the investigation, meaning she would work to ascertain the facts she needed. She was referring to the 5 W's.

Employers must not only demonstrate that they took action by conducting an investigation, but they also have to ensure that the investigation was "thorough." However, the term *thorough* is subjective. Consequently, whether or not the investigation was conducted thoroughly and in good faith would be determined on a case-by-case basis. The investigator has the discretion to decide what to include and what to omit, as well as which leads to follow and which to ignore. When these choices are made with the intention of protecting the company rather than the employee, it becomes apparent that the investigation is not conducted in good faith.

In my situation, the internal investigation was manipulated and used to falsify facts and evidence against me, creating false narratives. Many significant details of the unlawful conduct, which I had repeatedly stated and put in writing, were not included in the investigation. These are referred to as "material facts," and Jessica intentionally excluded them. She also chose what other employees to interview or not interview, what questions to ask or not ask, and I'm sure what to take note of or not. I had already experienced first-hand how manipulative she could be with statements and leading questions.

As an employee, you are not privy to these interviews or any other actions they take in the internal investigation. The process can take months and it had just begun. What they will state though is that the investigation is strictly on a need-to-know basis and that you should refrain from talking to others about it. It is stated in an attempt to silence you.

ANNUAL PERFORMANCE REVIEW AND ADVERSE ACTION

Shortly after the internal investigation started, I was scheduled for my annual performance review. Having been at HDR for three and a half years, this was my third review. My first two annual reviews were very positive; however, this last one was not.

As I mentioned before, Mitch told me he would be coming up from Fort Lauderdale to attend my review. At the time, I was unsure why he was making the trip. I also previously mentioned that, at times, Mitch and Laurence seemed to be used as scapegoats to shift blame away from Nick. This was one such instance.

Anticipating the possibility of them trying to fire me during the review, given everything that was happening, I asked Jessica to join the meeting as well. I still trusted her at this point, but it is crucial to remember that HR is not your friend. At the beginning of this meeting Jessica sat quietly for a change and observed. She may as well have had a bowl of popcorn with her.

The meeting was attended by Nick, Mitch, Jessica, and I. One of the first things they informed me of was the denial of the promotion to senior mechanical engineer—a position I had been told I met all the requirements for, that had been budgeted for, and that I was assured I would receive at the end of the year. Jessica had also previously assured me that my issues with Nick would not impact this promotion. Despite my heightened stress levels, I managed to maintain my focus on work and asked when the next opportunity for a promotion would arise. In hindsight, I realize I should have identified their actions as retaliation and called it out, but I simply wasn't aware of my rights. This was a clear instance of retaliation with the adverse action of failing to promote.

We each had a copy of the review in front of us. Written comments accused me of lacking empathy, hoarding work, and being confrontational—all of which were false narratives designed to undermine me. For instance, as I already noted, they had recently asked me during a weekly group call to delegate some work to Thomas. I took a moment in consideration and, once I had decided what tasks to delegate, they told me to think about it and reconsider. It was a setup, something I later expressed to Jessica, sharing that the incident felt like a trap. Their claim that I lack empathy is utterly ridiculous. I believe I possess a great deal of empathy, certainly far more than Nick. My empathy allows me to understand others' needs and take them into consideration. Their characterization of me as confrontational is also unwarranted. I prefer harmony and kindness in interactions, and any confrontations with Nick were a response to his degrading and offensive behavior towards me. While the review did briefly acknowledge my honesty and diligence, it was evident that they were now targeting my leadership and interpersonal skills, often referred to as "soft" skills.

Some additional feedback they provided verbally, but not in writing, pertained to the group performance evaluation conducted by Mitch on our team members. Initiated over a year before this, they had been monitoring us during that period. Nick and Mitch disclosed that the evaluation revealed I had been handling the majority of the workload, executing my tasks quickly, and producing high-quality results—the trifecta. This served as direct evidence contradicting Nick's discriminatory views against me. He had instigated the evaluation, believing I was inefficient and incapable, yet it unequivocally demonstrated that I was the most productive team member.

Even in the midst of this discussion, Nick, with a puzzled expression, managed to remark, "But your work is good." It was as if he couldn't believe the evidence glaringly before him. Later on, I emailed Mitch requesting to see the evaluation file. As a response, he proceeded to again gaslight me, insisting that his assessment was a "qualitative impression, not a quantitative comparison." Straight lies which he, as previously admitted, believed could not be proven. I wrote back to him asking, "How could it not be quantitative if a list of tasks, time, quality, and cost were compared for members of the group?" His follow up was "It cannot be quantitative, simply because it is not. You are berating me now. Please stop." Are you kidding me? Smoke, mirrors, and an insert of blame.

Annual performance reviews are frequently exploited for retaliatory purposes following complaints of discrimination or whistleblowing. The most glaringly ridiculous statement in this written review is them noting they had every intension to promote me this year based on confidence in my technical abilities and growth as a leader, however, subsequent to them making this decision to promote me I showed dissatisfaction in the current leadership... i.e. my whistleblowing. They basically spelled out whistleblowing retaliation.

As employees, you are entitled to respond in writing to these evaluations. They can be manipulated to blame the employee, thereby perpetuating the false narratives they've constructed, as was the case here. Following the meeting, I visited Jessica's office to express my concerns. I communicated that the accusations in my review were unfounded and that they made me uncomfortable. I worried that this review, compounded by Mitch forwarding the email including my confession of distrust in Nick to the Business Group, could jeopardize my long-term career at HDR. Jessica encouraged me to formally respond to the review comments, which I did. Intriguingly, during our conversation in her office, Jessica also suggested, "you should write a book"—and here we are.

BAD FAITH CAN BE TORTURE

The term 'torture' originates from the Latin word 'torquere', meaning to twist. Those who engage in retaliatory, bad faith actions, and intentional misconduct are essentially twisting reality and facts. These practices can be torturous for people, creating unsafe work environments and negatively impacting their mental and physical health. While employed by HDR, I suffered at times from loss of sleep and fluctuations in weight, severe headaches due to stress, and even skin rashes. I also had spurred anxiety as I questioned my future and at times depression when my complaints were blatantly ignored.

As humans, we have the right to security of person, defined by the United Nations Universal Declaration of Human Rights as the freedom from injury to the body and mind, ensuring our protection from physical and mental harm. This right is meant to shield individuals from intentional infliction of harm, both bodily and psychological. It is evident that Nick, with the complicity of others who enabled him and participated in the retaliation, were causing me harm and neglecting my complaints.

To me, at the very least, Jessica was just as responsible for enabling Nick. At times, it felt like they enjoyed what they were doing to me, as if seeing me in distress was somehow funny to them. By this point they had been plotting and lying for so long, the pain resided very deep within me.

NEGLIGENCE – INDIFFERENCE VS INTENTIONAL

Negligence is generally categorized under two umbrellas – "gross negligence" and "intentional misconduct". I had experienced both while at HDR. The difference between the two can be determined by view of intentions.

Gross negligence is when an individual or employer acts in such a way that they ignore your rights and basically turn a blind eye to what is happening, even though it's illegal. They act with a conscious disregard or indifference to the employee's life, safety, or rights. It is done with recklessness despite the obligatory wanting, or needing, in care.

Intentional misconduct, on the other hand, is when the employer knows what they are doing and are actively working to go against the employee's rights with malice. The employer knows what they are doing is wrong and they do it anyway, even with the understanding that there is a high probability of damage or injury arising from their actions to the employee or others, and even when the result IS damage or injury.

In my experience, HDR acted with negligence several times throughout my employment. They acted with gross negligence when they removed me from the bridge inspection tasks against Federal, State, and Local bridge inspection and safety regulations. They also acted with gross negligence by ignoring and dismissing my complaints of Nick's behaviors and the effects it was having on my health. While the ignoring and dismissing could be categorized as indifference, their actions of gaslighting and manipulating me throughout are acts of intentional misconduct, although they can try to continue to deflect this with more lies and denial. However, their actions of defrauding the FDOT are undeniably intentional as were their actions during the period after my whistleblowing and with regards to the failure to promote as well as the internal investigation. They acted with malice and caused me significant harm.

When employers see a potential liability, the employer may act with indifference or malicious intentions to silence, discredit, and harm the victim or complainant. Both versions of negligence can result in harm and both make the employer much more liable than they would be if they had just dealt with the situation in good faith.

FINDING THE LAW AND HOLDING THEM ACCOUNTABLE

At this point I could not fathom how any of this was happening. How did we go from my initial complaints in fall of 2020 about Nick yelling at me, belittling me, and threatening me, to now fall of 2022, receiving a poor performance review and being denied a promotion? It didn't add up but the stress of it all had. I realized I needed to seek advice elsewhere, which led me to consult with attorneys for a better understanding and more information.

What I discovered is that this happens frequently; it's systematic. The employers are aware of it, the lawyers are aware of it, but the employees are typically uninformed and blindsided. I shared my experiences of Nick's behavior over the years and my numerous complaints. Why weren't they helping me? The lawyers emphasized the importance of putting complaints in writing, not just verbalizing them.

We also discussed my whistleblowing related to the bridge inspection regulations. I recounted how Laurence designated to a client that I was the only one with the required training, yet they still prevented me from conducting the inspections, in turn performing them without a team leader present. We delved into the distinctions between fraud, gross mismanagement, and malfeasance.

A few things became clear in holding employers account-able. First, formal complaints in writing were necessary, uti-lizing unequivocal language that might be intimidating to say out loud. In addition, it was also the actions, or inactions, that follow complaints that were key. I was asked, "What did they do after your complaints?" Lastly, facts and evidence are required, including details such as dates, events, emails, text messages, documents, photos, etc., in which to create a time-line and prove what happened. Once again, the 5 W's por-traying their importance, without them your case against the employer becomes weak.

In reiteration, at a minimum, to hold them accountable:

1. Put things in writing, including complaints and instances to document misconduct.

2. Observe and take note of their actions and responses.

3. Record the facts and keep copies of documentation and evidence.

In my mind, I had already lodged a formal complaint. I had even requested a new supervisor, citing the harm Nick was causing me. However, one crucial detail I wasn't focusing on was the 'why'. I kept stating the facts, however, I hadn't been emphasizing that this was happening because I am a woman and Nick was sexist. I had repeatedly covered the who, what, where, and when, but didn't focus on the why. Jessica, at least, was fully aware of the reason and should have worked with reasonable care to protect me but didn't.

To force them to act and hold them accountable, I needed to finish the sentence. I had to put everything in writing, using terms like discrimination, harassment, and retaliation and note that it was based on a protected class. Eventually, I did just that. By this point though, HDR was negligent for so long already they chose to do everything they could to prevent me from moving forward with my complaints. They worked to destroy me.

It was also around this time that I decided to formally report the fraudulent activities externally, as HDR was gravely minimizing this issue too. I filed reports with both the State and Federal government representatives responsible for bridge inspections. Both confirmed that a qualified team leader is always required on site for bridge inspections, including movable bridge inspections.

FAMILY AND MEDICAL LEAVE ACT (FMLA)

So much had happened in just three and a half years at HDR. The more harassment and discrimination I endured from Nick, the more I complained. The more I complained, the more retaliation I faced. The more retaliation I faced, the more stressed and unhealthy I became. By the fall and winter of 2022, I had reached a level of stress that I had never experienced before. What I was put through has had effects on my emotional, mental, and physical health.

The headaches became so severe that they are difficult to explain and even difficult to find published research on. It felt like there was a ping-pong ball bouncing around in my head. As if the web of lies they had woven was beginning to unravel. The covert manipulations and gaslighting were finally coming to light, with an overwhelming amount, since it had been allowed to persist for far too long.

It is as if our brains are like a large ball of rubber band with each rubber band a piece of knowledge or experience. The rubber bands are neurons that transmit information from one cell to the next with chemicals called neurotransmitters. For years, HDR had been lying to me, manipulating and gaslighting me and now that I had begun to understand my rights, I could see the time more clearly. As though each individual rubber band formed by a lie started snapping in my head, I felt a literal sensation like a ping-pong ball bouncing around inside. I experienced what I believe was hyperactivity in my brain, so significant that it would eventually lead to what are called brain zaps and brain shivers among other symptoms that I will go into more detail on later.

Here anxiety and fear consumed me. What was happening to me? And what about my career and my work? What was I going to do now? The anxiety manifested not only in my mind but also throughout my body, extending into my arms—a strange sensation I had never experienced before either.

I had been pushed to a breaking point for sure. I couldn't take it any longer and was forced to request a leave of absence under the Family and Medical Leave Act (FMLA) in December. There are two types of FMLA leave: continuous and intermittent. I at first opted for intermittent leave partly in denial thinking that I could still contribute to the team by wrapping up ongoing tasks and be available to Jessica for the internal investigation. In hindsight, given my extreme stress levels, continuous leave would have been the more appropriate choice.

Jessica initially sent me the paperwork and had partly filled out some of the forms. Unsurprisingly, she did so in such a way to make it sound as if I was taking leave based on stress from my tasks and not Nick. I made sure to correct that before I sent her the completed forms. Another example of negligence. Somehow, I managed through my symptoms to continue to protect myself from their wrongdoing. I anticipated that this break from work would allow me time to recover, but what transpired was even worse. The blowback that was coming my way now would permeate my personal life.

DANGEROUS PRECEDENTS

My experiences at HDR and the research I have performed since have led me to a greater understanding that we need change to protect individuals in the workplace. Two main problems exist which set seriously dangerous precedents. First, employers and supervisors know the laws and discrimination rights better than those beneath them and have learned to work around the laws to the detriment of the employees. Secondly, the employer can do so with negligence and belief that they will get away with it even when the cost is the employees safety, health, or life.

While I was at HDR, I was oblivious to my rights and they took advantage of me because of it. When I tried to complain about the behavior, they worked intentionally to steer me away from the law. Instead of questioning if Nick treated me different based on my protected classes, they threw phrases at me like "Nick treats everyone the same"—although he most certainly did not. They have learned how to hide the ways workers are being treated differently due to a protected class by avoiding the use of facially discriminatory language or acts. If we dissect what HDR did, they were making Nick out to be just a bully to protect them from potential legal responsibility down the road while allowing me to suffer ongoing and growing stress. What would be worse though, someone who could be a threat to anyone or someone who is a threat to a particular group or class? It's dangerous precedent.

The instances that made me most upset or angry were when they were essentially taking from me. Nick's ways of degrading and diminishing me took away from my value and worth as an individual. My responses came from attempting to protect myself and not wanting him to continue to hurt me. There were also the threats to my job and conspiring to "get rid of me" that created emotional reactions of fear that they could take away not only my current job but also my career and livelihood.

The negligence, whether indifferent or intentional, that employers will engage in for efforts to protect their reputation is the second dangerous precedent. As employers have much more power and resources than the employees it has become completely unfair and justice is out of reach. Lies and cover ups can be worse than the original acts themselves. But worst of all, those with high power have the ability to use this eras data and technology against the victims. If employers and entities were to continue to act this way, we will be left with more and more suffering on top of distributive corruption without accountability. Corruption is like a virus that spreads, the only anti-virus is the truth.

8

FILING CHARGES AND HOW BAD IT WOULD GET

LAWSUITS AND COMPANY REPUTATION

Lawsuits are necessary in serious matters when the accused refuses to take accountability for wrongful actions. The case can be brought to court to be seen and evaluated by a judge and jury. Discrimination and the Labor and Employment laws around these types of situations are a big deal. Employers tend to be highly cautious regarding various forms of discrimination, as legal action and a guilty verdict during legal proceedings could significantly tarnish the company's reputation.

Discrimination cases can severely damage a company's public image. This is why HR's primary goal is to safeguard the company and its reputation. A company may go to great lengths to protect its image and prevent a legal battle, particularly if they are aware of their own unlawful misconduct. This is an expression of corporate power, and the lengths they will go to might vary depending on the potential damage to the company's reputation.

AGENCY FILED COMPLAINTS AND CHARGES

This next decision should be approached with careful consideration. If the situation deteriorates to the point where filing charges becomes necessary, a good first step is to identify your relevant agency. A universally accessible option, due to its federal jurisdiction, is to schedule an appointment with the EEOC to file a complaint, which you can initiate online. Alternatively, you can seek out state or local government agencies, keeping in mind that the location of your workplace can influence which agency has jurisdiction over your case.

Bear in mind that these agencies often have a backlog of cases, as many others are also attempting to file complaints. Consequently, it might take months before you secure an appointment, which is for filing your complaints—not your charges. It's also important to be aware that there are statutes of limitations for filing charges, and they can vary depending on the agency you choose. During the complaint process, you may work with an agency investigator, who will also expect detailed summaries of events, including the 5 W's, making your preparation of documentation and a timeline of events indispensable. Alternatively, you can proceed to filing charges through a lawyer. Either way, the charges have to be filed within the statute of limitations for your rights to be preserved. While working with the agency investigator or a lawyer, you can determine what claims you have to individual charges or even multiple allegations for each charge.

From this point forward, continue to document diligently. If you decide to proceed with charges against your employer, be prepared for the possibility of severe blowback and further retaliation, which could extend into your personal life. They may leverage their corporate power to target you on a personal level in extreme ways. Being a single woman living alone made me a more vulnerable target. It is difficult to comprehend unless you have experienced it firsthand.

For these reasons, I believe we need to better hold employers accountable on our own terms when it comes to discrimination, fraud, or other unlawful practices within the workplace. Squashing it as soon as possible is the best chance for resolution with limited damages. This starts with proper documentation and reporting. That is the first line of defense in holding them accountable. To do this, you would first need to be aware of and understand your employment rights and the relevant laws. While everyone has the right to pursue legal action, based on my experience, this should be considered a last resort—an option when the company's conduct is so egregious that you are left with no other choice, as I was.

INTERNAL INVESTIGATION CONTINUES

While I began my pursuit towards legal action, the company continued their internal investigation. Jessica reached out a few times during this process to schedule some follow ups, continuing to interview me as required.

It was about a month after my whistleblowing when I reported the wrongdoing to HR and office management. Along with these reports made to both Jessica and Noreen, I again asked to be transferred out from under Nick's supervision. However, it wasn't until about two months later during the internal investigation that HDR finally made the transfer. Unfortunately, they transferred me now from under Nick's supervision to temporarily under Mitch's supervision although I had disclosed he as well had retaliated against me after the whistleblowing. Even during this official internal investigation, they continued to act in bad faith and negligence. I was naïve to assume they wouldn't.

A question I raised during this period was for Jessica to confirm what was sent to the Business Class Group for them to deny the promotion. As I stated, Mitch had told me that he sent them the email where I noted having a lack of trust in Nick. In the meeting when I first brought it up, Jessica said she would look into it and in a later meeting when I attempted to follow up about it, she refused to confirm.

Jessica did disclose a few names of individuals she was planning on interviewing for the investigation. I noted that those individuals were people who would favor Nick and requested some others I would have liked her to also talk to. Her response was that she may not be able to get to everyone and that it depended on where the facts lead her. Toxic employers acting in negligence are not going to be looking for the truth, they are going to work to document what they believe will protect the company—even with deliberate lies. They can come up with excuses as to why you were not fit for the job or search for things to use against you, however, these could be pretextual.

Most significantly, from the internal investigation came a "confirmation letter" which noted a list of complaints I had made. Similar to a written performance review, you have the right to respond or comment to a document like this. After Jessica sent this to me I found many errors and misrepresentations throughout. I have addressed and commented on this confirmation letter multiple times now, having gained a better understanding of the law over time and finding new points to make. Rather than conducting a thorough investigation, they clearly used the process to bolster their case against me.

During this period, I also found that others were asking me relevant and leading questions. For example, I had an employee from another office reach out by instant message. They asked me several questions about an old task. Recall and memory are important for credibility. In addition, he started to go on saying that he was ready to retire although he was not. It seemed to me he was fishing to get me to say something similar but I was nowhere near ready to retire. I loved my work.

DOING MY OWN RESEARCH AND INVESTIGATING

But while HDR was investigating internally, I was also dealing with induced stress and anxiety while searching for answers and gathering more information. Anxiety often stems from the fear of the unknown, so I worked to learn and understand what had really happened, why it happened, and what my next steps should be in moving forward. If I tried to remain idle, it only made my head hurt more, as though a force inside my mind was pushing me to act.

What if they decided to give me back the promotion? What if they fired them? What if they fired me?? The lawyer I spoke with said he would bet on them doing nothing. I was also advised that if they attempted to fire me and I was forced to sign something, I should sign and mark next to my signature: "Acknowledging receipt, but not in agreement." However, that scenario never occurred.

While searching for answers on how to approach the problem at hand, I was only beginning to grasp the full extent of the situation. All I wanted was to feel safe, but I couldn't. They had allowed everything to continue for too long. The harm done to me had a profound effect on my well-being.

While it's probably not wise to threaten legal action outright, it is in your best interest to demonstrate an understanding of your rights and explicitly document in writing how they were violated. After my initial attorney consultation and throughout the internal investigation, I noted—both verbally and in writing, to the best of my understanding at the time—the ways in which HDR had violated my rights.

One of the most critical decisions is selecting the right lawyer or attorney. Attorneys typically work on either a contingency or a retainer basis. A retainer requires an upfront payment for legal services, while a contingency arrangement means the attorney assumes the risk and gets paid only from future settlements or financial awards. I intended to have legal representation, and by this time, I had engaged an attorney on contingency. This attorney filed the charge on my behalf and issued an initial demand letter to HDR, which was necessary to preserve evidence. Although I had already initiated an EEOC complaint on my own, this attorney improperly filed a duplicate charge with both the EEOC and the Florida Commission on Human Relations (FCHR). Because of this, my case was investigated by the FCHR rather than the EEOC, as I had intended.

One unresolved question remains: Was this attorney truly working in my best interest, or were they influenced to sabotage my case? Ultimately, I was forced to proceed legally through self-representation, also known as pro se litigation.

In legal proceedings, there are options for informal dispute resolution, primarily mediation and arbitration. Arbitration is typically something an employee must agree to via contract, usually at the start of employment. It is a final process and generally favors the employer. Mediation, on the other hand, can be offered multiple times throughout the legal process, starting as early as after filing an initial complaint with a state or federal agency. While I opted for mediation multiple times, HDR repeatedly declined. Unless a judge orders mediation, it is necessary that both parties agree to it, for it to take place.

THE POSITION STATEMENT AND FALSE NARRATIVES

Once the agency investigation begins, after you have made your initial complaints, the employer has an opportunity to submit what is called a "Position Statement". Since my experience is limited to Florida, I can't speak to what happens in other states, but what HDR did here was glaringly neglect-ful, unethical, and wrong. Their Position Statement was sub-mitted as two PDF documents totaling 169 pages, filled with misleading information, false narratives, and outright lies.

Jessica also submitted a sworn affidavit that completely ignored the years of complaints I had made. Instead, it focused only on the period following my whistleblowing, failing to acknowledge any wrongdoing and portraying me as the problem. She specifically claimed that she had been coaching me during that time—although this so-called "coaching" was nothing more than manipulation, pushing me to apologize to Nick in an email. Once I realized what Jessica had been doing—manipulating and gaslighting me—the emotional distress intensified.

Is it okay to react? I believe the most important thing is to be honest and to respond in a way that aligns with what you are experiencing. If necessary, you may have to explain your actions, so bear that in mind when considering how to proceed. Throughout this entire dreadful experience, I have worked hard to be honest and tell the truth. HDR, on the other hand, did the opposite. If you didn't handle a situation perfectly due to stress they caused, that's understandable—it may not be ideal, but it's okay. Just don't lie about it.

When I first skimmed through the documents, I was actu-ally calmer than I expected to be. Employers and their law-yers use systematic tactics to deter employees from moving forward. It feels as if they construct walls and barricades to block your path—obstacles you then have to break through and tear down. Even though their Position Statement was lengthy and packed with misleading information, I remained calm because I could clearly see the lies—and with every false statement, I knew how to counter it with the truth.

The most alarming part of their Position Statement was a section containing notes Jessica took from an interview with Thomas. While many of the falsehoods in their report involved misrepresentations—such as omitting key or material facts, misattributing dates, or misstating names—Thomas's statements were the most shocking of all.

This was the same guy who, near the start of my employment, tried to coerce me into having sexual relations with Nick—a guy who was frequently and blatantly inappropriate. And now, he was fabricating outright lies about me. It was the section that caused the most significant pause in my initial review.

DEFAMATION AND PRIVILEGE

What is defamation? Surprisingly, the concept is not as straightforward as one might think. In my opinion, the definition has been stretched, overstepped, and is now out of control. As Mitch once said to me, "You can just lie."

Defamation is supposed to protect us from false statements—whether spoken (slander) or written (libel). It is meant to safeguard our reputations from being damaged by untruths. However, instead of serving its intended purpose, employers have weaponized falsehoods to further harm victims, destroy reputations, and evade accountability.

There are two primary defenses to defamation: qualified privilege and absolute privilege.

Qualified privilege applies when someone makes a false statement, but it was made with good intentions—typically to warn or protect someone who is perceived as a threat to themselves or others.

Absolute privilege is meant to be a complete defense against accusations of defamation, protecting government officials and individuals involved in judicial proceedings from liability for false statements.

But both of these defenses create more problems.

In the context of employment disputes, employers have abused qualified privilege with reckless disregard for the truth. Instead of acting in good faith, they act in bad faith, maliciously pushing employees to their breaking point. When that level of stress is finally reached, they quickly point the finger and say "She's the problem" like they did with me. Instead of stepping in to protect me, my employer neglected their responsibilities and later lied to cover it up, making the entire situation so much worse.

I can understand why absolute privilege originated. Government officials and those involved in legal proceedings often speak publicly and work on complex legal cases, so they need protection from errors. However, intentional lies and honest mistakes are not the same.

What disturbs me the most about defamation laws is that even witnesses in legal proceedings are protected under absolute privilege. While this is meant to encourage honest testimony, I find it unsettling that it has instead

FILING CHARGES AND HOW BAD IT WOULD GET

created an environment where people believe they can lie without consequences—simply because it's difficult to prove otherwise. Technically, intentionally lying under oath is perjury, a criminal offense. But how often is perjury actually prosecuted?

MY REBUTTAL

After taking the time to review the Position Statement, responding to it seemed like a daunting task. As with any such task, you can always break it down into parts to make it more manageable. The response to the Position Statement is called a Rebuttal.

It was around this time that my engagement with my attorney became rocky. They had already acted oddly in some ways, including telling me to go out and buy something nice for myself or saying that this case would never go to trial. It was now just two days before the deadline for the Rebuttal, and my lawyer abruptly terminated our engagement. It was devastating, as I was abandoned at a crucial time.

Having already been struggling mentally and even physically, this only added to the stress. Feelings of hopelessness can arise and become overwhelming. You have to find the will to steer yourself away from bad thoughts, but admittedly, it is not easy. I feared not being able to solve this on my own and, in turn, never getting accountability. I'm an engineer— solving problems is what I do. But I am not a lawyer, and I had only just begun to learn my employment rights.

203

I found hope from an unexpected source—an old piece of notebook paper from high school that I had saved. It was an exercise where everyone wrote their name at the top of a paper, which then got passed around the room for classmates to add notes about that person. It may seem ridiculous, but at that moment, I needed to hear those comments. They included: "Independent woman. Very secure about herself," "Great funny girl! Always smiling and laughing," "Good person," "Such a sweet spot under all that toughness" and "Determined."

Life can make us forget who we are. Those notes helped remind me of myself, at a time when I was so beat down, degraded, and feeling defeated. I have always been one to stand up and speak out when someone is being hurt, even when it's a stranger. During this entire period, I have been fighting and researching not only for myself but for all the others I know are hurting in or from similar situations. That has been a bigger driver and motivator for me than just fighting for myself. I could not give up. The words snapped me out of it and gave me the confidence I needed to keep moving forward.

So here I was, with no lawyer and a huge task in front of me. Thankfully, the investigative agency granted an extension on the deadline. Again, not moving forward made my symptoms worse, so I had no choice but to get to work on the Rebuttal. Line by line, sentence by sentence, point by point, and even with the footnotes, I argued against their statements with the truth.

IN GOD I TRUST

Believing in God is a personal choice, one that not everyone feels compelled to make. It's a complex decision, given the various beliefs one might hold. Some may choose to believe in the universe, while others may not subscribe to any belief in a higher power, attributing life's events to chance. Ultimately, the choice is a personal one. However, I do believe that the values associated with belief in a greater power are significant. From childhood, I was taught to do the right thing with the thought that God is watching, God knows.

The Orthodox Christian faith, which I was raised with, differs in ways from the Catholic faith, including some traditions. Most Orthodox churches, for example, do not offer confession or even have confessionals. During this particularly challenging period in my life, I found solace in turning to God. In the Orthodox tradition, we don't typically ask God for forgiveness in the way some other denominations might, as we acknowledge our human nature and capacity for error. The teachings I received encouraged accountability and moving forward after making amends for mistakes. In our prayers, we more often seek salvation and deliverance from evil, asking for assistance when someone is ill, when we need help, or when we need strength and guidance in our journey. We may reserve our prayers for moments when we are out of options and in need of divine intervention. During my ordeal, I prayed and called out to God for the exposure of truth and deliverance from the transgressions surrounding me.

In addition to this, I leaned into my family's following of a particular saint named Saint Paraskeva or Saint Petka. Paraskeva means preparation and Petka means fifth or Friday, and the names have been used interchangeably. As a result, I found there are three distinguishable saints who have been mixed up and confused over centuries, all represented by the same name. One from Rome, another from the Balkans, and the third from Turkey, they each have distinct symbolism. Saint Paraskeva from Rome is shown in icons or images to hold a bowl with a pair of eyes and said to cure the blind or heal ailments of the eyes, which could be taken as a source of exposing truth. The second from the Balkans represents kindness and compassion for others and worked to help the sick and needy. And last but not least, the third from Turkey was rich with love and devoted herself to family. Growing up, we had icons of the version with Saint Paraskeva holding the eyes within our home, however, because of the confusion between the three we celebrated her on the day that represented the one from the Balkans. Though now, I can clearly see and feel connected to all three.

I also understand the perspective of trusting in the universe, viewing life as a series of events and choices that lead us to our current situations. This perspective led me to consider that perhaps my past had been preparing me for these current challenges and something greater.

9

CONSPIRACY OR NOT CONSPIRACY, THAT IS THE QUESTION

MENTAL HEALTH CRISIS

It is upsetting to say, there is a mental health crisis in the U.S. and areas around the world. But what is the cause? One cause that has in the past been overlooked or downplayed are the effects of ill intensions and betrayal. Is it the individual that is claimed to have mental health issues who is at fault? Or is it that something or someone drove them to it? Is it possible that in the U.S. there is a serious problem of lack of respect and dignity? Is it that those in power believe and have been taught through previous experiences that they can get away with things that they shouldn't?

Suicide rates in the U.S. are also very high. In March of 2021, a man named Evan Seyfired, a manager at a Kroger supermarket in Ohio, took his own life after being harassed for roughly a year by two other store managers. Reports say that they also began to interfere with his personal life by allegedly hacking into his electronics and sending him obscene messages. I have also been subjected to such actions as a result of my reporting on HDR's misconduct and have deep sympathy for his family and loved ones who lost him to such evil and malice.

If you cause extended stress in someone's life, the blame should not be directed at the person who has become stressed, but instead at those who caused it. What if a mob bands together to trap a person, leading and driving them closer and closer to the edge of a cliff? Out of fear and stress, the person may start to react, they may lose their own mental focus or control. But what if the mob closes in on them further, to the point that there is no more room to spare and the individual loses the ground beneath them, falling off the cliff? This is not a suicide. If the mob didn't touch the person though, is it murder? It is prolonged malice and ill intensions that can drive a person off a cliff, whether literally or figuratively. These types of suicides can occur because the victim can't take the stress anymore, in the moment they have lost mental control by the suffering they are put through, and avoidably decide this is a way out.

Then there are the mass shooting rates in the U.S. which are reportedly higher than other countries. Some may think it is a gun control problem, but Scandinavian countries also have high gun ownership rates, although without the mass shooting problems. Where is our issue coming from? What is driving this crisis?

In the fall of 2023, there was a mass shooting in Maine acted out by Robert Card, killing 18 people, followed by taking his own life. Reports noted that he was a military reservist and had previously received a number of awards for his service. In the summer leading up to the mass shooting, Robert was institutionalized for two weeks in a mental health facility, with claims he had been "behaving erratically" and "hearing voices". The story with regards to his motives were unclear. Sadly, these types of deadly event happen resulting in the loss of innocent lives, including adults and children.

The reports become insistently directed at labeling the shooter with mental health issues and covering mainly the events of the shooting and the woeful effects on the community it occurred in. The stories hardly cover what happened leading up to the shootings. Did Robert really need to be committed to a mental health facility or was it that someone with power manipulated the situation, discrediting Robert, and being the true cause of the mass shooting and consequent loss of innocent lives?

What could drive someone to the degree of believing your only choice is homicide? Relentless and out of control anger, which can be triggered by betrayal and injustice, seeking revenge? Think of an instance, such as your significant other cheated on you with your best friend. How would that make you feel? Hurt, angry, betrayed. Now, imagine that on a larger scale of injustice where someone has manipulated the truth, worked to take away your career, and possibly ruin the rest of your life by tarnishing your reputation, and having you wrongfully labeled with mental health issues. This type of thing happens, it is what HDR attempted to do to me. I had some bad thought as a result, but I would never act on them.

Thoughts and actions are not the same thing. We can choose to control ourselves not to act on extreme thoughts.

In both of these very unfortunate types of circumstances, suicide or revenge killing, the individuals are being pushed to believing they have no other option, but there is always another option. You can choose not to hurt yourself or hurt others. You can choose to fight within the confines of the law. There is a phrase or old legal adage that I have come across a few times this year, "If you have the law on your side, pound the law. If you have the facts on your side, pound the facts. If you have neither on your side, pound the table." I think a more helpful version of this would be as follows: "Learn the law, so you can pound it. Document the facts, so you can pound those too. Use your voice, to pound the table."

ILLEGAL STALKING AND SURVEILLANCE

In addition to our rights in the workplace, we have a right to privacy. When a company becomes adversarial, it can resemble a toxic relationship breakup, akin to an ex-partner resorting to stalking. This can escalate to illegal stalking and surveillance. I anticipated finding relief from the workplace misconduct while on FMLA. However, I instead found myself embroiled in the toughest experience of my life.

I use the term "toughest" deliberately because, although I needed time to recover, I was forced to withstand even more and continue fighting to protect myself without breaking. All I wanted was to feel safe, yet I found myself in a relentless battle for my own protection. I don't know how this experience will affect me long term.

The limits of your privacy come into question here, along with the extent of their intrusion. While public spaces do not afford much privacy, this is considered the norm. However, when interference or close monitoring begins to impact your life negatively, it crosses a line into harassment and retaliation, which is unequivocally unacceptable.

In quantum physics, there is a renowned experiment known as the double-slit experiment, devised to explore the behaviors of subatomic particles. Researchers used particles like light photons or electrons, and shot them at a barrier with two slits. Behind this barrier was another wall. Initially, they observed that the particles displayed wave-like behavior. Curious to gather more data on the particle-wave movements, they implemented monitoring devices. Astonishingly, this act of observation altered the particles' behavior; they ceased exhibiting wave-like properties and instead behaved as independent particles.

Drawing a parallel, when filing a lawsuit, you may experience extreme measures of illegal monitoring and surveillance. This may include hiring private investigators or other parties to stalk and surveil the complaining employee, or ex-employee. If monitoring subatomic particles at such a minute level can influence their behavior, it raises the question: how could this level of surveillance not affect the employees under scrutiny? Without doubt, I was stalked and surveilled during my time at HDR and since. While I cannot pinpoint exactly when the monitoring began, I suspect that even before it registered consciously, it was affecting me subconsciously.

It was as though they were present in my life, observing from a distance without revealing themselves. However, at a certain point, they escalated their actions, invading my personal space and deliberately causing disruptions in my life. Some individuals involved in such activities can display sadistic and sociopathic tendencies, deriving pleasure from witnessing others in pain without regard for the rights of the person being stalked. It's unsettling. They can surveil at anytime, anywhere, and employ various tactics to get inside your head.

One instance that depicts the extent of the situation in public occurred outside a Walmart. I was at home one evening, cooking crepes, when I realized the need for a new pan as the crepes were sticking. Deciding to purchase a new nonstick frying pan, I headed to Walmart. I picked out a pan, paid at the self-checkout—opting not to bag it since it was just the pan—and exited the store, holding the pan by the handle at my side, somewhat like a tennis racket. As I walked through the parking lot, I suddenly heard footsteps rushing towards me. A woman ran up behind me and said, "It looks like you want to hit someone with that pan," into my ear, and then ran off. It was an attempt to antagonize me to chase after her. This unsettling encounter was neither the first nor the last of its kind since the escalation of issues at work.

Gradually, I began feeling more and more vulnerable, as if I couldn't do anything or go anywhere without wondering if I was being watched, or if someone would attempt to mess with me. Even my own home's back patio felt off-limits due to concerns of surveillance. I kept my blinds closed and minimized my time outside, even going so far as to install additional security measures, including outdoor cameras. Initially, this constant vigilance made me question my sanity, due to its unexpected nature. At first, thoughts of it just being a coincidence arise, but after how many incidents can you conclude that this can no longer be chance? Similar to Nick's sexual harassment, I didn't want this to be true, and many times I tried to ignore it. The time he inappropriately hugged me in the doorway of his hotel room and my attempting to pull away as he continued to grasp onto and caress my arm comes to mind. That along with my swift walk down the hallway shaking my head afterwards, not knowing what to do. It is crucial to remain confident in your perceptions and control reactions to provocations.

Much like addressing misconduct in the workplace requires speaking out, the same applies to situations infringing on personal life. Consequently, I extended my documentation to encompass these personal incidents as well. However, gathering the 5 W's in these scenarios proved challenging, as I couldn't identify the people involved. Despite this, I was able to collect some information.

Another concern was the potential hacking of my electronics, as they seemed intent on discerning my plans. Even Jessica, during the last few months of my employment, tried to probe, asking, "What's your plan?" At times, it felt as though they anticipated my next move or referenced things I was doing in private. How was this happening? Had they tapped my phone, hacked my laptop, or bugged my home? I couldn't be certain, but if they had, it would constitute a severe invasion of privacy and illegal discovery. Can you imagine that it could be even worse than that?

The entire ordeal resembled a high-stakes game of chicken with attempts to intimidate into abandoning the lawsuit. They appeared willing to go to any lengths to achieve this. From my perspective, retreat wasn't an option; surrendering after everything I had went through seemed unthinkable, the result of which could equate to never fully recovering. To endure this period, I tightened my budget, cutting non-essential expenses, and continued to invest time in researching and learning about the legal process to ensure I reached the end.

DAVIS V. HDR

During my employment with HDR, there was a separate lawsuit filed against them on privacy rights named *Davis v. HDR*. This case was initiated in the fall of 2021 and concluded at the beginning of 2023. Corporations are constantly involved in lawsuits, and although there may not be a direct correlation to my case, I believe it is worth noting. Overlapping this with my timeline of events, the case began around the time of the sexual assault and my removal from the Virginia bridge inspection tasks and ended near the conclusion of my employment.

The case involved a class action lawsuit alleging violations of the Electronic Communications Privacy Act (ECPA). The allegations stated that HDR had unlawfully monitored and collected data from two private Facebook groups. Although the groups were private, the judge ruled in HDR's favor, determining that the information they obtained was not protected because there was no agreement of confidentiality or privacy among the group members. The judge also stated that the communications within the groups were "readily accessible to the public."

Similar to my presence at Walmart or any other place considered public, the law may unfortunately consider this legally obtained information. HDR's general argument was that they were merely observing what was already publicly available. But at what point does this level of surveillance go too far? As I already noted, merely being observed with the intent of data collection can alter a person's behavior. How is this acceptable?

Another major distinction that should be made here is the intent behind this intrusion of privacy. Why was HDR collecting data from these groups? The two groups HDR infiltrated were both based in Arizona: Ahwatukee411 and Protecting Arizona's Resources & Children (PARC). Ahwatukee411 is a community forum for residents of the Ahwatukee Foothills area, discussing local issues. PARC was formed to protest the construction of a highway cutting through the Moahdak Do'ag Mountains. An article by *The Architect's Newspaper*, titled "HDR Responds to Reports of Surveilling Clients' Opponents", referred to these groups as activists.

Apparently, HDR has a team called STRATA dedicated to this type of electronic data monitoring. This team provides 24/7 social media monitoring, intended to identify, analyze, and mitigate risks for clients, stakeholders, and—undoubtedly—HDR itself. In the *Davis v. HDR* case, HDR monitored these groups because their activism opposed projects HDR was working on. Another article by *Vice*, titled "A Company That Designs Jails

Is Spying on Activists Who Oppose Them," described HDR's STRATA team as an example of what some scholars call "corporate counterinsurgency." The article also noted, "When social movements threaten profits and political agendas, corporations and the government sometimes work side by side to neutralize those who oppose controversial projects."

Here's the thing about me: for the majority of my life, I have stayed off social media. I have typically been very private. I don't usually post pictures of myself or share specific details about my life. I generally found social media to be a waste of time and preferred to focus on other things. The one platform I engaged with the most was LinkedIn, mainly because it was centered around my career and networking.

However, throughout this experience, there have been multiple instances where individuals have questionably encouraged and manipulated me into posting and commenting more on public social platforms—both on major social media sites and within my community. It felt as though they were actively working to force information out of me, similar to how police officers pressure suspects into making false confessions. That is not okay.

Over the past two years, my increased engagement on LinkedIn has been completely out of character for me. This change was driven by the distress caused, along with my desire to help others understand their rights so they don't have to endure what I went through. LinkedIn also became an outlet for my emotions—a place where I could share my experience and not feel completely alone while the stalking, antagonization, and surveillance pushed me away from other social norms.

PERPETUATING FALSE NARRATIVES

Not only might there be attempts to run you off with stalking, surveillance, harassment, and continued retaliation in your personal life to intimidate and antagonize you, but they may also work to perpetuate the false narratives they created in the workplace. They might believe that this will strengthen the case they are building against you. It is a horrifying experience. If they choose, they could attack every aspect of your life to discredit you.

The false narratives may vary, depending on what they can find in your life to exploit. They may look for attributes to twist into patterns, or take misfortunes from your life and use them against you in their favor. As I have disclosed, I am a single woman living alone which introduced vulnerability and was exploited. These people can begin to try taking everything away from you that could help you. For a long time, I kept distance and separated from others to not only protect myself but them as well. I even question if I have had relationships under false pretenses, whether professional, friendly, or romantic. Something I may never get truthful answers to.

I have mentioned my membership in Toastmasters, which again brings the matter of privacy into question. Toastmasters is a professional organization geared towards building public speaking and leadership skills, and technically, investigators could follow me to my meetings. Observing from a distance is one thing, but interference is quite another. I initially joined to help strengthen my voice in better standing up to Nick, and was a part of my club for approximately two and a half years. Initially, I was shy, but I developed good relationships with the other club members. Over time, I was nominated for and accepted three different officer or leadership roles: sergeant at arms, vice president of education, and finally, president of the club. However, as issues at work escalated, I noticed a change in how I was treated within the club, as if I was being undermined and discredited. I also found myself catching questions during Table Topics sessions that seemed eerily related to my work troubles.

For those unfamiliar with Toastmasters, Table Topics involves spontaneously responding to a question in a one-to-two-minute answer to enhance impromptu speaking skills. It felt as though the club was being used to trap me with more leading questions. One question I received, for example, was "How much money do you need in your bank account, Diana?" It is an odd question to get and at the time still in denial about what was happening I would just answer the questions, responding to this one in summary with "Enough so that I feel safe."

If HDR is willing to intrude on privacy the way they did in the *Davis v. HDR* case, what would make anyone think that they wouldn't also intrude on someone's privacy in person. While Toastmasters played a large role in strengthening my voice, the mounting stress from my workplace and uncertainties about the club being compromised eventually led me to leave. If HDR had intended to undermine my ability to demonstrate positive leadership skills, infiltrating my Toastmasters club would be a subtle and effective strategy.

CONSPIRACY OR NOT CONSPIRACY

We have already touched on conspiracy relating to HDR's actions within the workplace, but does it go further than that? A conspiracy occurs when two or more individuals or entities agree to partake in unlawful or harmful actions. In such cases, those assisting in the misconduct may be coerced, influenced, or bribed—effectively being weaponized against you. The challenge, however, is that you are usually not privy to this agreement. Consequently, without the potential aid of a subpoena, it may be impossible to confirm its existence.

Is it conceivable that HDR would go to such lengths? Would they employ coercion or influence to compel others to assist in retaliating against me? If it was happening at Toastmasters, was there an actual agreement or financial exchange involved, or were they simply undermining and discrediting me? They might resort to slander or other discrediting tactics to persuade parties to act on their behalf. But what kinds of individuals or entities might they enlist?

Certain locations are inevitable parts of daily life. Naturally, your own neighborhood could be a target. Since it is your place of residence, your neighbors might find themselves coerced. What about supermarkets? You need to eat, and if you're saving money by not dining out, grocery stores become frequent destinations—making them another possible avenue for coercion and influence.

Other places to consider are those where people seek relief or solace. For instance, some might attend church for prayer. In such a sanctified space, one would expect to feel safe and unthreatened. However, during this time, I read from a catechism that even addresses the presence of this in the church and suggested to get out of there.

Then, there is bribery—a scenario where the employer pays people to act against you. The most likely candidates for this kind of bribery are so-called expert witnesses, which could include professionals like therapists, doctors, or even local law enforcement. And what about lawyers? Is it possible for an attorney—someone who is supposed to be advocating for you—to be bribed to work against you instead? It has indeed happened before; some lawyers get bought out.

If all these conspiratorial acts begin unfolding, panic is almost inevitable. You might even begin to fear for your life, questioning what lines wouldn't be crossed after having already crossed so many. Whistleblowers often find themselves at serious risk.

Despite these challenges, the priority remains self-protection and meticulous documentation. While the instinct may be to go on the defensive or even flee, it is possible to confront and navigate through the fear. A constructive approach involves not just focusing on what is happening to you, but also closely observing their actions.

What are they saying?

What are they doing?

How are they approaching me?

Recalling the 5 W's becomes crucial yet again. Document everything—take videos, pictures, license plates—and build your evidence.

LAW ENFORCEMENT

Living in Florida definitely has its perks, and the beaches are one of them. I chose to settle down in the Tampa Bay area and bought a townhouse in a suburban neighborhood on the outskirts of Saint Petersburg, about 15 to 20 minutes from the beaches. The specific area I live in borders Pinellas Park and has an Eastern European community, which I have greatly appreciated. It has provided the comfort of finding familiar items from my North Macedonian cultural cuisine, such as ajvar, a vegetable spread made from roasted red peppers and eggplants.

Unfortunately, after the whistleblowing, this community has shifted from comforting to very uncomfortable. Not knowing what else to do about the stalking incidents, I eventually turned to my local police, the Pinellas Park Police Department, for help. Naturally, one might assume that this would be a good step toward seeking assistance and protection.

When I first started reaching out to the police department about the issues in my personal life, I was still employed with HDR and on leave under FMLA. One day, I spoke with Officer Miller outside the police station. While I tried to explain what I was experiencing, he seemed more focused on me than on my concerns. He asked me questions about whether I was planning on leaving HDR, questions about my family, and whether I was seeing medical professionals. I would soon decide that I could no longer continue working at HDR, but that was none of his business—nor were some of the other things he was asking. His priority should have been helping me with the stalking, but instead, it felt like he was trying to shift the blame onto me. It reminded me of how Jessica used to treat me.

A few months later, after more incidents, I decided to email Officer Miller about specific instances where I wanted to request store security footage. One of the incidents I mentioned was the encounter with the woman at Walmart, noting that she "spoke right into my ear." Later, the officer took my words and altered them in a separate email to a store representative, misrepresenting my statement to say that I reported someone had "whispered" in my ear.

The very next day, this same officer showed up at my house with a mental health crisis representative, attempting to Baker Act me. In Florida, the Baker Act allows for detaining someone if they pose a threat to themselves or others. I was neither—but HDR was a threat to me, and Officer Miller was failing to do his job to serve and protect me. Not only was he failing, but he was actively working against me by altering my statement and using it as a pretext to force me into a mental institution, much like what they did to Robert Card in Maine. It was an abuse of privilege. I began to question whether the police department had also been influenced.

A few months later, I was still being stalked and harassed. One day at the supermarket, a woman approached me holding a package of pig ears. Since the end of my employment with HDR, I had been carefully watching my spending, buying sales. As I stood in the meat section looking for the best deal, this woman suddenly walked up to me with the raw pig ears and said, "Here, buy these. These are cheap. Cook with these."

It wasn't friendly. There was aggression in her tone, and

as she moved toward me, I instinctively took a few steps back. But she kept pressing forward. Finally, I said, "I see them," then turned around and began to walk away.

Once again, I reached out to the police department to request the security footage—this time, going over Officer Miller's head to his sergeant. The response I received was even less cooperative. The sergeant refused to investigate, effectively obstructing justice.

Even further, this police department actively worked to defame me and coerce others against me. I requested the police body cam footage from my interactions with this department. In one particular report, the officer deliberately and clearly defamed me by falsely stating to others that I had made statements that I did not. He also made slanderous comments and then coerced the individuals to make complaints about me to my HOA, in attempts to have me kicked out of my home.

This police department has worked to obstruct justice, defame me, and coerce others against me. These actions amount to official misconduct and negligence. As if what I endured at work wasn't already enough, for about six months after my constructive dismissal, I had to deal with law enforcement actively working against me, further exacerbating the stress caused by HDR.

The thought of a cop working against you is terrifying. They are supposed to serve and protect you. But, now, I was worried that if HDR or someone else tried to physically hurt or kill me in retaliation, I would have no protection. Would that also get covered up?

10

RESPONSIBILITY AND OBLIGATION

MAKING A FINAL CHOICE

For me, loyalty played a role in the timing and manner of my decisions to speak up and the decision to leave HDR. That along with being unaware of my rights, were the two most significant factors. In life, sometime we stay loyal to people longer than we should. Maybe it is because we are attached to an idea of what could be. Now, I was in the process of grieving losing the idea of my career. The more something or someone means to you the more substantial the grief is.

Overall, I didn't want to see Nick lose his job. I recognized the opportunity to learn from him, especially regarding the technical aspects of movable bridges. At the same time, I found it so hard to work under him as it had an enormous effect on me emotionally, mentally, and even physically causing a hindrance in concentration, loss of confidence in myself, and exhaustion. We attempted to establish boundaries many times, but realistically the relationship was doomed from nearly the beginning starting with Nick slipping his hand down the back of the chair and resting it next to my butt.

Protecting oneself becomes challenging when trying to maintain loyalty to someone or to a company who doesn't provide the respect and decency we all have a right to as humans. This dilemma became most apparent leading up to and during the course of the whistleblowing. When the internal investigation began, I was still in denial hoping that everything would turn out okay. I repeatedly inquired about the potential impact on my future with the company. Until almost the very end, I expressed fondness for HDR, despite everything. The situation had been unbearable for roughly three years by this point, much longer than I should have allowed.

As the investigation progressed, it became clear that they were not acting in good faith. Jessica sent me the confirmation letter, including a list of my complaints. During the investigation, several instances of the sexual harassment started pouring out of me. Prior to consulting with a lawyer, I had assumed that the law would protect me simply by stating that my boss was causing me harm. Why wouldn't it?

It was February 6, 2023 when HDR had submitted its Position Statement to the investigative agency. Ten days later, I had a virtual meeting with Jessica and another HR manager to discuss the conclusion to their fact finding and internal investigation. I was told they "were not able to substantiate discrimination, harassment, or retaliation." An apparent standard response in these types of toxic workplace investigations. They also completely ignored the whistleblowing and fraudulent activities and barely included Mitch and Laurence in the investigation although I clearly stated in writing that I believed all three had retaliated against me.

They also provided a short list of reasons for denying my promotion, all of which stemmed from Nick's discriminatory treatment towards me. They used these incidents to continue labeling me as having poor leadership skills, when it was Nick who had the deficit of leadership and management skills along with a lack of professionalism to refrain from harassing behaviors and sexual intensions. When Jessica conveyed this list, she rattled them off so quickly I could barely jot them down quick enough in my notes. I requested the list via email and rebutted each point in writing.

I was still expecting them to do the right thing, yet here they were again downplaying the situation. HR is not your friend so don't expect them to act like they are. Remember, they are more like "Company Resources" than "Human Resources". I responded that I still believed I was discriminated, harassed, and retaliated against by Nick, Mitch, and Laurence. They then brought up my FMLA and asked me when I would be ready to rejoin the team. I immediately noted feeling it would not be appropriate to have to continue to work with this same management.

At this point, my FMLA leave was nearing its end. I see now that it should have been an easy decision to get out of there but it was as if I was stuck holding on to a future I had envisioned for myself. The company said it geared its values around people and creating quality work, which very much aligned with my own core values. They promoted this by noting a focus on behaving respectfully and professionally as well as supporting the employees to direct their own career paths, becoming technically advanced in the field of their choice. However, Nick did not encompass these values when it came to me and neither did any of the people who joined in to protect the company instead of properly dealing with the problem. It was abuse followed with a neglectful cover up.

What really shed light on it all was going through the Position Statement and producing the Rebuttal. The lawyer I had engaged with questionably terminated our agreement on February 20th when the Rebuttal was due two days later, leaving me in a terrible position. With the extensions granted I was able to submit the Rebuttal about a month later. Being forced to prepare the Rebuttal on my own, by the end it became so obvious, I could no longer ignore or see past it. There was too much manipulation and too many lies. I now saw and felt the betrayal and injustice of it all more than ever, three and a half years' worth summed up into two large documents. With everything they had done, I finally realized that there was no way I could stay. I was forced to resign effective March 24, 2023, a decision that amounted to constructive dismissal.

Here is a summary of that timeline for a better overview:

February 6, 2023 - HDR had submitted its Position Statement.

February 16, 2023 - Conclusion to HDR internal investigation.

February 20, 2023 - Lawyer terminated our agreement.

February 22, 2023 - Rebuttal was originally due.

This was not laid out by chance. It seems as if it was designed intentionally from a playbook. If it has happened to me, how many others has this happened to? What is it doing to the overall health of the people in this country? Is it only this bad in Florida?

THE WAITING

Yet another challenging aspect of this process was the waiting. It takes months for investigative agencies to review the materials provided to them. They may also conduct interviews with the individuals involved. In the end, you may receive one of three determinations: yes, there were unlawful actions; no, there were no unlawful actions; or they could not determine whether unlawful actions occurred. The EEOC allows for all three of these outcomes, but the FCHR only provides a yes or no determination, which can limit your rights to due process by forcing you into an administrative hearing instead. Having a complaint reviewed by the FCHR therefore works in favor of the employer. Regardless of what the agency concludes, your life can end up on hold while you wait.

One thing I am profoundly grateful for is having my dog Cooper by my side during this time. Being forced into isolation, his presence was a great comfort. He also gave me tasks to do, fulfilling my need for activity and engagement, as I greatly missed my work. The movable bridge work I performed brought me joy, and completing tasks left me proud of my accomplishments. My work was a significant part of my life.

Cooper has also given me a different perspective on this ordeal. I have worried about the toll this experience has taken on him. It has not been fair to either of us. But this perspective also made me realize the trauma that can be inflicted on the significant others and even children of those going through similar situations. They aren't just harming the victims they seek to silence—they are also hurting their families.

The emotional, mental, and physical distress caused by this kind of ordeal should not be taken lightly. At times, I found myself driving aimlessly just so I could scream at the top of my lungs, blasting music to drown it out. I'd return feeling much better. Due to the stalking, I was forced to withdraw from socializing. But it is vital to find ways to release the stress and tension these situations create.

Documenting, journaling, and writing played crucial roles in keeping me focused and moving forward. Many times, I felt as though they would rather I die than let the truth come out. These activities helped me envision a possible positive outcome—one centered on doing what's right and sharing the lessons I've learned. Surprisingly, I developed new interests in history, law, and politics—a major shift, as these topics had never intrigued me before.

Despite the relentless attempts to intimidate, dissuade, and shake me off course, I held on for dear life, refusing to let go.

Eventually, I received a no-cause determination from the FCHR, due to HDR's misleading statements and misrepresentation of the truth. Even the agency investigator's summary included inaccurate information. Instead of opting for a forced administrative hearing, which would have limited my options, I chose to have the EEOC conduct a substantial weight review—which meant even more waiting.

RIGHT TO SUE AND LITIGATION

The first stage of waiting ends when you receive the "Right to Sue" letter from the agency that investigated your complaint. This letter allows you to begin the litigation process by filing a Complaint with a court and initiating a lawsuit. The individual filing the Complaint is called the "Plaintiff," and the opposing party is the "Defendant." There are strict time limits, known as statutes of limitations, which are critical and can expire if not met. Up until this point, everything is considered pre-litigation.

It's worth noting that AI and chatbots can now help navigate these processes and quickly reference relevant legal information.

Before filing, you must choose the proper venue, which determines jurisdiction over your case. This is typically based on where the alleged unlawful actions occurred, where the corporation conducts business, and where the individual defendants reside. You must also decide whether to file in state or federal court. Based on my limited understanding at the time, I chose to proceed through state court.

In my area, there are two primary counties: Pinellas County (covering Saint Petersburg) and Hillsborough County (covering Tampa). I initially filed in Pinellas County, but the case was later transferred to Hillsborough County.

After filing a Complaint, you are required to serve it on the Defendants through a certified process server. The costs associated with this process were approximately

$400 for filing the Complaint and around $100 to serve it. Corporations have a registered agent designated to receive legal documents and forward them to the company, preventing them from evading service.

Once the corporation and its lawyers get involved, they will likely try to stall or delay proceedings as much as possible, hoping you will run out of money, become exhausted, or simply give up. The first move to expect from them is the filing of a "Motion to Dismiss." There are various types of motions that can be used throughout litigation.

The Motion to Dismiss, at least the first time around, felt like a gut punch that knocked the wind out of me. It reminded me of the dismissals I used to receive from Jessica and others when I reported misconduct at work, but now on a much larger scale. Just as you have the right in the workplace to respond to their weak excuses, in litigation, you have the option to file a "Response in Opposition" to any motion they submit. If your side files a motion, you may also have the opportunity to file a Reply to the Response, depending on the court's procedural rules.

Other types of motions that can be filed include:

- Motion to Compel: A request to force the opposing party to respond to the complaint or comply with a discovery request.

- Motion to Stay: A request to pause the proceedings for various reasons.

- Motion for Injunctive Relief: A request to prevent the opposing party from taking certain actions before the case is resolved.

Once all responses and replies are submitted, the judge will review the facts and arguments presented and issue a decision or ruling—either in a written order or first through a court hearing. However, there doesn't seem to be strict rules on how long judges have to process filings, meaning more waiting should be expected.

Another essential aspect of litigation is the discovery process, which mainly includes:

- Interrogatories: Written questions that the opposing party must answer under oath.

- Requests for Admission: Statements that the other party must either admit or deny.

- Requests for Production: Formal requests for documents, records, or other evidence.

In my opinion, interrogatories are the most useful tool for the Plaintiff. They allow you to directly question the Defendant and hopefully obtain the answers you've been searching for. Some questions I feared asking—such as whether others in my life had been influenced or whether they had intruded into my home or other property—were the very questions that, when asked properly, could offer protection. Once you put it in writing and demand an answer, they should stop.

I also sent them a cease-and-desist letter.

Requests for admission are direct statements that you ask the opposing party to admit to. Examples include:

- "Admit that Plaintiff was sexually harassed by an HDR employee."

- "Admit that HDR performed multiple inspections without a qualified Team Leader on site, against Federal, State, and Local regulations."

- "Admit that Plaintiff met the requirements of the Senior Mechanical Position."

- "Admit that Plaintiff was given a poor performance review after she made protected complaints."

The standard limit appears to be around 30 interrogatories and 30 requests for admission, though this may vary from court to court.

Production is the stage of the process in which parties must "produce" documents or other tangible forms of evidence, such as virtual call records, emails, instant messages, and more. Additionally, you have the option of filing for subpoenas. To me, however, requesting subpoenas feels like digging into others' private matters, and that doesn't sit well with me.

MEDICAL CARE, OR LACK THEREOF

Throughout this experience, there have been ups and downs. There were periods when things seemed to settle down, allowing me to focus on the case and writing this book. However, there were also times when my symptoms became extreme. It was unfair, to say the least, that while I was working to seek accountability and justice, I also had to contend with serious concerns about my health. I can't say exactly how long the sensations in my brain lasted, but it went on for months. Remarkably, when I sought medical care, the medical professionals failed to provide proper care.

Not long after filing the Complaint, I began noticing new sensations. These felt very different from what I had previously described as a ping-pong ball bouncing around or rubber bands snapping in my head. What I was experiencing now felt more like electrical sensations, and there were two distinct variations. It was difficult to find documentation about this online, but eventually, I came across the terms "brain zaps" and "brain shivers." The limited information on these terms associates the symptoms with either periods of severe stress and anxiety or withdrawal from antidepressants.

Initially, I believed that my symptoms—including the ping-pong ball sensations, brain zaps, and brain shivers— were a direct result of the stress caused by HDR. But there was more to it than that. It was now January 2024, nearly a year after my constructive dismissal and all the retaliation extended into my personal life. Following filing the lawsuit, I finally felt safe enough to seek medical care.

I first considered getting brain scans, so I attempted to visit an imaging center in Saint Petersburg. However, they turned me away, stating that I needed a prescription from a doctor. In an effort to avoid potential surveillance, I had placed my phone on airplane mode and parked several blocks away, walking to the imaging center. When the representative there suggested a nearby clinic called Allcare Clinic, I eagerly asked for directions. It was a few blocks north, so I walked there, mostly taking side streets.

When I arrived, I waited to speak with the front desk, hoping to be seen that day. However, I was told that I would need to return the following day for an appointment with Dr. John Ross.

The next day, I arrived for my appointment and was eventually directed to an exam room. Soon after, Dr. Ross entered. I had informed his office and him personally about my whistleblowing and discrimination lawsuit, even providing them with a copy of the complaint for their records. I also explained that this situation had been ongoing for quite some time and had caused me a great deal of stress.

I tried to describe my symptoms, but at the time, I still didn't know the proper terms. Instead, I mainly described them as headaches and sensations in my brain. I explained that I believed these symptoms were related to the gaslighting, manipulation, and lies that HDR had subjected me to. I told him that it had all started after I learned about discrimination laws. It felt as though HDR had created a false reality, and when that false reality suddenly and drastically collapsed, my body responded in ways I was struggling to understand.

He specifically asked a few questions relating to migraines and I denied having experienced any of those symptoms. I was not experiencing, nor had I ever experienced, migraines. I went on to explain the severity of the symptoms I had been experiencing.

In addition to the brain sensations, there were instances when I felt my head move involuntarily, which was very alarming. The movements were mostly slow, forward, and downward motions. Another concerning symptom was an instance of temporary blurred vision in my right eye. It appeared as a partial dark spot. I went to the mirror to see if something was caught in my eyelashes or if there was something in my eye, but there was nothing. At the time, I decided to turn everything off—the lights, the TV—and I laid on the couch in darkness and silence for a few hours until the dark spot went away.

During the visit, the doctor seemed very sympathetic and understanding. I trusted him. I also confided that I had been dealing with stress-induced rashes, which would flare up under the pressure of everything I was going through. He prescribed me an ointment for that. He also gave me a prescription for an MRI scan, which I had done later that day. These were clearly not normal symptoms or reactions, and I needed to know what was causing them.

Additionally, they scheduled me for a follow-up visit, which I didn't think much of at the time. When I returned for the follow-up, I requested the doctor's notes from my first visit so I could review them. When in a situation like this, it's important to check all documented notes to review and retain a copy of what is there.

Back in the exam room, while waiting for the doctor, I briefly skimmed over the notes and reviewed the lab results from blood work I had also done since my last visit. I was still trusting of Dr. Ross, and once he started the exam, I willingly answered all his questions. We discussed my brain symptoms again, and I went into greater detail about the electrical charge sensations. I explained how it felt like a distinct electrical charge or current, typically originating from the back of my head and traveling over the top or around the side toward my right eye. He assured me that once the legal proceedings were over, these symptoms should go away.

We also went over my lab results, which, thankfully, were generally normal. He then began asking me questions related to a woman's exam, and without giving it much thought, I answered freely.

Later that evening, when I got home, I started reviewing the doctor's notes I had obtained from the first visit. I was both shocked and disturbed by what I read. He had clearly falsified my records and documented that I had reported symptoms of migraines. It was very obviously not an error. In multiple places, statements were attributed to me that I had never made. These included that I reported experiencing auras with flickering and shimmering of lights, that the auras reached complete resolution prior to the onset of headaches, that I reported having a family history of migraines, and that the migraines improved with Excedrin Migraine. I didn't say any of that! He completely fabricated all of it, and more than once during that visit, tried to convince me to buy Excedrin Migraine.

The next day, I went back and requested the notes from my second visit. This time, he didn't lie about my reports, but instead, he documented a Well Woman Exam. I was never asked if I wanted a Well Woman Exam, and had I been asked, I would have directly refused it.

In strict legal terms, in the first visit, he falsified my medical records. In the second visit, he performed and documented an exam without my consent.

Ultimately, I found the terms online that best described my symptoms around this time: brain zaps and brain shivers.

Brain zaps are the sensations I described earlier, which feel quite literally like electrical charges or currents zipping across parts of your brain.

Brain shivers are different—they feel more like your brain is vibrating or jiggling, similar to how your body shivers when you are cold. Some shivers also caused a slight twisting or torsional effect.

I also noticed a repeated pattern in my symptoms: induced stress, followed by brain zaps and/or shivers when the stress subsided, then heartburn. In my entire life, I only recall one specific instance of heartburn before this period, and that was from some potent garlic knots.

I believe what happened was that my brain became over-stimulated—causing the ping-pong ball and snapping sensations—which led to swelling. That swelling then caused tension headaches, not migraines. The pressure from this swelling and tension triggered my other symptoms, including involuntary movements, a brief episode of blurred vision, and the brain zaps and shivers.

As noted earlier, the limited information about brain zaps and shivers that is available online links them to antidepressant withdrawal, but I was not on any medication. Some sources also mention that these symptoms can result from extreme stress, fear, and anxiety.

This creates a "chicken or the egg" dilemma—which came first? Did HDR's actions against me within the workplace cause the stress, or was the stress intentionally induced allowing for a means to blame me and label me as the problem?

It seems that hyperactivity could be more damaging to a person's health than the brain zaps or shivers, given that it causes a significant release of hormones and neurotransmitters that can impact other organs and bodily functions. However, all of these symptoms were alarming.

To me, both zaps and shivers felt like residual effects from the hyperactivity or overstimulation—as if my brain was trying to settle itself back down, cognitively with the zaps and spatially with the shivers.

I now also believe that brain zaps and brain shivers are linked to two separate components of the brain: zaps are related to the electrical component, and shivers are related to the magnetic component. Imagine two magnets vibrating next to each other—that's what a brain shiver feels like.

One might wonder why I am sharing all of this. My answer is simple: to work toward a future where people are not denied proper healthcare and to prevent falsehoods from continuing. "Havana Syndrome" is a term often used to describe some of these symptoms or similar ones. By definition, a syndrome is a set of symptoms or conditions that occur together, suggesting the presence of a disease or an increased risk of developing one. By calling it a syndrome and implying it is a disease, they could also attempt to prescribe medication. I will never take prescription medication for what they have put me through.

More importantly, labeling this a syndrome is just another form of victim-blaming. It is not the victim's fault. It is purely an attack on the victim through means of advanced neuro-technology. This advanced technology can interact with the brain and nervous system, including methods and devices for monitoring and manipulating neural activity. This is what caused my symptoms above. I will continue to explain and prove that my symptoms were intentionally induced as part of an effort to prevent my success in the legal proceedings against HDR.

DATA PRIVACY AND HACKING

In today's world, there is an enormous amount of information online—some of it accurate, and some of it misinformation or disinformation. Every individual or entity that contributes to social media or creates content online adds to this vast pool of information. You don't just create a profile of yourself on one particular site—there is an umbrella profile that extends across every site, application, and piece of data you generate through data cross-referencing. This information can be analyzed, monitored, or even weaponized against you—as seen in what HDR has been doing with their STRATA team.

Those with the means and resources can acquire your data—and you may not even realize how much data you produce through the applications and devices you use daily. For a long time, I couldn't understand how I was being followed. The stress and symptoms caused by my headaches and the effects of weaponized neurotechnology made it difficult to process everything at times.

It took me over a year after my constructive dismissal to realize that my vehicle itself could be used as a tracking device. I had already considered the possibility of a physical tracking device being attached and had searched for one, without finding anything. However, modern vehicles come equipped with built-in GPS and can transmit location data. Once I made the connection, I checked my vehicle manual to see how to disable it. I discovered that removing the micro-SD card disables the GPS.

Some data could be used against you, while other data could work in your favor. It can be overwhelming when you don't know the legal process or what a corporation or other entities might do with your information. I have leaned toward assuming that data will be useful and have made an effort to keep all subscriptions and relevant records intact. Legally, you should not be deleting anything that could be relevant anyway. However, they can also target your credit cards with fraudulent charges, potentially causing autopay subscriptions to be canceled if you aren't staying on top of updating your payment information. Privacy settings are often not straight forward, and many people don't pay close attention to them.

In addition to data monitoring, there is also the threat of

being hacked. Your accounts, phone, laptop, and other electronic devices could be compromised. Never in my life would I ever consider attempting to hack someone, but it is a serious threat. Around the same time, I disabled my vehicle's GPS, I also put my phone on lockdown mode. Much earlier, I had already started using a VPN and other security measures, including changing passwords frequently and freezing my credit across all three credit bureaus. I took every step I could think of to prevent them from gaining access.

MIND DATA PRIVACY AND NEUROTECHNOLOGY LAWS

Unfortunately, although we theoretically have a right to privacy, advancements in neurotechnology can overstep these rights. How do you defend yourself from something that can't even be seen? Or something that you don't realize is happening? How do you stop them from getting in when you are not even aware of the possibility of what they are capable of? And how do you protect yourself when they are always a step ahead of you because they are listening to your thoughts?

You may not be aware of this, but neurotechnology can listen in on your inner thoughts and collect brain data. Neurotechnology is advanced technology that is developed to understand and access the brain. Imagine the position you'd be in if someone could hear every thought you had. Most of our thinking, decision-making, and understanding isn't even conscious—it's subconscious. It is estimated that only 5% to 10% of human thought is conscious, while the remaining 90%

to 95% is subconscious. If that's the case, then those who can listen in might know you better than you know yourself. On the other hand, brain data might also be misinterpreted—leading to incorrect assumptions about a person's thoughts, emotions, or intentions. From what I've observed, thoughts and actions don't always, or even usually, align.

However, collected brain data may be much more than just your thoughts. The brain acts as the body's processing and control center, similar to a control room of a movable bridge, only far more complex. It is connected to the rest of your body through the nervous system, effectively controlling every function. What is frightening is that if someone can control your brain, they can control you. While I cannot be certain how it happened, how long it has been going on, or even who might have explicitly orchestrated it, I know that certain experiences I've had were not under my control. My empathetic nature can't help but wonder how many others have suffered or how many lives have been lost as a result.

Evan Seyfried?

Robert Card and his innocent victims?

I have gone from being an introvert who didn't mind working in a corner for privacy to realizing that privacy can be completely taken away—even the privacy of your own thoughts. It has felt like being in isolation, while simultaneously having no privacy at all. For a while, it took all my energy, leaving me unable to recharge the way an introvert normally would when alone. It is an experience that goes beyond typical feelings of violation or betrayal. You start replaying the past, realizing embarrassing things you did when you thought you were alone. Even daily routines and natural bodily functions become a source of anxiety that you can't hide from. You can't hide from anything.

Legislators around the world are finally beginning to consider implementing human rights protections for mind data and neurotechnology. Chile was the first country to enact laws protecting the privacy of the mind and mental integrity, known as "neurorights". These laws were introduced in 2021, following a case brought by former Senator Guido Girardi against Emotiv, an American neurotechnology company. The case centered around Emotiv's devices, which are capable of monitoring brainwaves and collecting neurodata. Girardi argued that the company was improperly storing and processing brain data without adequate legal protections. This case set a global precedent, leading Chile to design new legislation protecting brain data and brain activity—the first of its kind in the world.

The most alarming issue here is the lack of consent. These technologies can currently be used without the knowledge or consent of individuals whose minds are being intruded upon. Chile's neurorights laws establish protections against unauthorized brain data collection from neurotechnology devices (such as Emotiv's) and require explicit, authorized consent before such data can be accessed or used.

The next issue these neurorights address relates to mind data collection, data security, and exploitation. This part of the legislation safeguards against mind data being hacked, misused, or commercially exploited without consent. In such cases, they are stealing your thoughts with the intention to profit from them. It's basically human trafficking and slavery. Chile's law protects against unauthorized access with actions of reidentification, use for advertising, or for the purposes of surveillance.

Lastly, and arguably most importantly, these neurorights protect mental integrity, prohibiting manipulation or interference with brain activity to alter a person's thoughts, emotions, or cognitive processes without consent. This protection is meant to guard against cognitive and psychological manipulation.

While the first two issues focus on what can be done simply by listening to someone's thoughts and how those thoughts can be used as stolen property, the protection of mental integrity is focused on how external influence can be imposed on an individual instead.

There is also the Neurorights Foundation, based in the United States, which originated at Columbia University. The Foundation works to promote innovation while protecting human rights, and aims to ensure the ethical development of neurotechnology. The Foundation played a role in developing Chile's neurorights legislation and continues to advocate for neurorights protections in other countries and at the United Nations, helping to lead the global movement. Within the United States, Colorado is currently the only state to have included "neural data" under the definition of sensitive data in the Colorado Privacy Act.

For me, it has been over two years since hyperactivity was induced in my brain. I have documented several occurrences since then, and although I did not recognize the pattern at first, these occurrences pointedly aligned with the timing of the lawsuit against HDR.

THE "JUSTICE" SYSTEM

At the start of my long haul through the "justice" system, I was still incredibly naïve. Optimistically, I believed that getting my foot in the door of the court system would provide relief and justice. However, more than a year later, after enduring further cruel and extreme punishment simply for expecting accountability, I am now forever changed.

The justice system is used and manipulated by corporations like HDR and their lawyers as a tool to avoid accountability, exploiting court rules and procedures to delay cases and deny accusations, regardless of their fundamental authenticity and basis in fact. But I suffered much more than just this along the way.

Delay and deny, over and over—that is the strategy. They use various legal loopholes in the court's rules and procedures, hoping they will never be forced to accept accountability. Some would rather see the employee end up homeless or even risk their life than admit to misconduct. As if their reputation is worth more than a former employee's life—a perilous comparison.

At this point, my case against HDR has been put through three separate courts, beginning with the state court system in Pinellas County. I specifically chose this court because I had the impression that state courts were more favorable to a Plaintiff than federal courts. Additionally, I had other reasons—mainly the stalking, continued retaliation in my neighborhood, and police misconduct. My assigned judge was Judge Thomas Ramsberger.

After filing the legal Complaint, HDR's lawyers filed a Motion to Dismiss which hit me hard when it arrived. A hearing was set to address both HDR's Motion to Dismiss and the Motion for Injunctive Relief that I had also filed. Injunctive relief is a court order demanding that certain actions such as retaliation stop. This was—and has remained—the most important issue for me. The retaliation in my personal life and everything I was being put through was causing further harm, worsening my distress, and preventing me from healing. I mean it sincerely when I say it feels they would rather see me dead than accept accountability.

At the start of the hearing, I was generally calm and hopeful. Judge Ramsberger first addressed the Motion to Dismiss. When choosing a venue (the specific court where the case is filed), there are specific legal criteria that must be met to ensure that the venue is appropriate. HDR's Motion to Dismiss was based on an alleged lack of clarity in my Complaint regarding why Pinellas County was a suitable venue. Technically, Pinellas County was a proper venue because:

1. HDR had conducted work in Pinellas County related to the case

2. Nick, one of the Defendants involved, resided in Pinellas County

However, I had not explicitly outlined this in the Complaint. I could have and would have filed an Amended Complaint to correct this, but the judge denied my request to do so. Instead, he granted HDR's motion, and my case was transferred to Hillsborough County. With this single ruling, I lost three months—just for the case to be re-filed and considered active within the new court.

More notably, not only did he deny me the opportunity to amend my Complaint, but he also denied me the chance to be heard at all on my request for injunctive relief.

At this moment during the hearing, I suddenly developed a very sharp, distressing headache, making it incredibly difficult to think clearly. It wasn't until months later, after seeing the patterns emerge, that I realized this was due to interference with neurotechnology.

Moreover, it was around this time that I visited Allcare Clinic and had my medical records falsified. Justifiably upset—having reported the symptoms I did, only for them to be wrongfully and negligently documented as migraines—I once again sought out legal assistance, this time trying for a personal injury lawyer. While I repeatedly found that attorneys would not take my case, I also repeatedly found that I learned things just from speaking with them. Unconscionably, even with serious brain symptoms, I still could not find an attorney willing to represent me. One lawyer, however, did offer some insight about Judge Ramsberger. He noted that Ramsberger was the most difficult judge in the area and that the judges in Hillsborough County were generally better than those in Pinellas County.

As time went on, I continued to suffer from retaliation, but I also continued to learn. I have filed multiple separate lawsuits. I became like Oprah—"You get a lawsuit, and you get a lawsuit, and you get a lawsuit!"

The second lawsuit I filed was against Allcare Clinic and Dr. Ross for medical malpractice, including the falsification of my medical records and for performing and documenting a Well Woman Exam without my consent. The third lawsuit was filed against the Pinellas Park Police Department for official misconduct and negligence, including their acts of falsifying records, defamation, obstruction of justice, coercion, and failure to perform their duty to serve and protect me, a citizen in their jurisdiction.

All three lawsuits were filed in Pinellas County. And all three lawsuits were assigned to Judge Ramsberger. Coincidence? I think not.

With the medical malpractice lawsuit, I found that the attorneys were even worse than HDR's lawyers. While HDR's lawyers worked to exploit the justice system through delays and denials, the lawyers representing Allcare Clinic were malicious: they didn't just defend their client, they actively manipulated the system to take even more from me. Both the attorneys and Judge Ramsberger acted this way in this case. Judges are not supposed to impede pro se litigants—in fact, they are supposed to allow for leniency on technical matters. However, in this case, the opposite happened. The judge dismissed my case and awarded attorney's fees to the clinic's lawyers—meaning I was forced to pay for their legal defense of a doctor and clinic that falsified my medical records, all while I was seeking medical care for brain-related symptoms. Are you kidding me?! I was forced to pay about $10,000.

At the time, it stung—it felt morally wrong—but I also knew that one day, I would publish this book and include this experience. And in the end, that makes it worth it. My point is not to sow doubt in the justice system or specifically damage Judge Ramsberger's reputation. My point is to expose the injustice in the system in hopes of improving it. The same way I did with HDR. If no one complains, if no one holds these people accountable, it will only continue—and it will get worse. I believe we are at a point right now where we actually have a real opportunity for transformative, positive change.

While I later filed the lawsuit against the police department, I never even served the complaint on them—out of fear of Judge Ramsberger. It is also worth noting that, during the medical malpractice lawsuit, I had two separate hearings with Judge Ramsberger. Both of those times, I did not experience any onset of headaches or induced confusion, as I did during the HDR case hearings. In fact, my mind was relatively clear and without pain during those hearings.

But returning to the HDR case: after the case was transferred to Hillsborough County, they filed another Motion to Dismiss. This time, their attorneys targeted my allegations and causes of action. The original Complaint contained six separate counts:

- Discrimination based on gender (including harassment and sexual harassment)
- Retaliation
- Hostile work environment
- Conspiracy
- Whistleblowing retaliation
- Defamation

Similar to my Rebuttal during the agency investigation, I responded to all their baseless arguments with reason and the truth.

The hearing for this Motion to Dismiss was scheduled a few months out, for July 2024, due to concerns regarding my health. Additionally, I had re-filed the Motion for Injunctive Relief, and also submitted a Motion to Compel, hoping the court would force HDR to respond to the Complaint. These two hearings were scheduled separately for May 2024.

The first hearing addressed the Motion for Injunctive Relief. As I previously stated, this was the most critical motion for me, as it could have improved my living situation by stopping the retaliation and stalking, in hopes of allowing me to seek proper medical care. At this time, I was still unaware of the capabilities of neurotechnology and the fact that it was being used against me. Unlike in the Allcare Clinic case, my mind was noticeably affected during this hearing. I suffered from induced disorientation, confusion, a severe headache, and an overly emotional response. I am sure this was observable, and it may have influenced the judge's decision to deny the order for injunctive relief. Thankfully, these hearings were virtual, and shortly after the hearing ended, the induced sensations stopped—almost as if they had been turned off like a switch. However, my fears remained: without injunctive relief, HDR and the retaliation remained an ongoing threat to me.

With unmet concerns for my health, I sought out another doctor's office.

What I really needed was to see a neurologist, but I could not risk them falsely reporting a condition that could lead to coercive and involuntary detainment. Instead, I sought out another general practitioner, primarily to obtain proper documentation in doctor's notes and to undergo additional lab work. I had been experiencing pains in my sides near my kidney region, which caused additional concern. Fortunately, all lab work was reported within normal limits.

Although neurotechnology can target the mind, it can also affect other parts of the body. Another induced symptom I suffered was changes in my heart rate and blood pressure. This doctor was the first to document it, noting that my blood pressure was significantly elevated to about 140/100—which was abnormal for me. This concerned me enough to purchase a blood pressure monitor.

A day or two before the Motion to Compel hearing, my symptoms escalated again—I experienced persistent headaches, confusion, and emotional distress. I checked my blood pressure—it had now spiked to 155/110. I had no choice but to request that the court reschedule the hearing with a continuance. This hearing was my best chance at ending the hell I was being put through—and they knew it.

If the motion had been granted, HDR would have been forced to answer the Complaint. Instead, I was attacked and tortured to prevent that from happening. The patterns in my symptoms became more apparent and defined. I could see at times my heart rate would go up, my blood pressure would spike, and the headache and confusion would begin. Once things finally settled down, often while watching TV or when I'd start to relax and let the mind wander, I would experience the brain zaps or shivers, and later heartburn.

The next hearing was for the Motion to Dismiss and was scheduled for a Monday. Some more happened, though, over that weekend. By this point, I did not know what to expect and believed I had to do more to protect myself and the case. So that Saturday, I began to release things online. I posted compiled documentation of what I had been put through at HDR, with the police, and with Allcare Clinic. This decision was extremely difficult—I didn't want things to get worse. Even more so, difficulty arose from additional mind manipulation causing pain, such that I was left resting my head on my arm, sprawled out on my desk, knowing that I needed to post but feeling so much resistance in doing so.

This resistance is like wading through water or, worse, through mud. I believe part of what caused much of my pain was my refusal to turn in the direction of the manipulation and mind control. It is like they can try to force you down the wrong direction, but I knew I had to go forward, and heading forward was met with induced pain and confusion. Pushing through it, I eventually posted.

The next day, a Sunday, the judge filed an order canceling the hearing along with granting the Motion to Dismiss but without prejudice, meaning I could file an Amended Complaint. He also provided a clear and detailed list of items that needed clarification, which helped in my amendments. Over this time and through lessons from the Allcare Clinic case, I continued to learn and gain understanding of the laws and rules. So, in addition to making adjustments based on the judge's order, I added three new counts, including negligence, assault, and violation of the Fair Labor Standards Act for the forced work without pay. I have filed evidence with the courts along with affidavits and declarations under oath justifying all of the now nine total counts.

Unfortunately, instead of addressing and responding to the filed Amended Complaint, HDR and their lawyers made their next move, causing further delays by transferring the case from state court to federal court. There are differences between the court systems, mainly their own sets of rules and procedures I had to learn anew. It was exhausting.

Now August, this was also around the time that I started to realize what was really happening: that there were forces being imposed on me and I could not control them—neuro-technology. It is not a "syndrome," it is an attack. Basically, the justice system can be rigged against individuals. What kind of "justice" is this?

DATA COLLECTION VS MIND CONTROL

Considering the idea that your thoughts can be listened to having already been addressed, we can focus more on the difference between data collection versus mind control. Within engineering and technology, often data sets, algorithms, and various systems are comprised of inputs and outputs. In terms of neurotechnology and mind data, I consider the collection of this data to be an output and the mind control or manipulation to be an input.

Unless a path is specifically and sufficiently designated to allow for travel in only one direction, it can always be traveled in either direction. Take, for example, a one-way street. It is technically designated for travel in only one direction; however, it is still possible for someone to travel down the street in the wrong direction, although it would be illegal. If your mind data is currently being collected by someone—or possibly by AI as an output—then they would also have access to your brain for input and control.

As previously noted, others can collect and exploit your mind data for their gain. In my case, the mind data was being used to manipulate me in various ways. They can track your routines, plans to leave the house, and other details about what you are doing. This data can be used against you in stalking, antagonization, and manipulation making comments about things they couldn't have known otherwise. The intention is to cause anxiety, doubt, and distress.

But they can also use your thoughts as their own or learn from you, like a focus group that would normally be paid compensation, only they are doing it without pay. They are essentially stealing and using the individual like a slave, only the shackles are not visible, and the individual is not even aware of it. It is modern-day slavery and human trafficking.

However, in addition to this, they can also impose things such as suppression and target someone to prevent them from speaking up against illegal actions. Similar to coding, it is as if they can add and remove code from your mind. Think of Newton's Third Law—for every action, there is an equal and opposite reaction. The mind coding can be like a wall. If someone were to push against a wall and then that wall suddenly disappeared, the person would end up falling flat on their face. It felt this way many times, where the control would stop or start again in discrete instances.

It is inhumane, and if they did it to me, they can do it to anyone. It is a human right to have security of person. But with these new technologies, we now need security measures for intrusion on our thoughts and minds.

AI AND THE NEED FOR REGULATIONS

If knowledge is power and power yields wealth, then we need to share the wealth through the transfer of knowledge. Even people like Nick will try to limit the information they share to hold on to power, but that doesn't help the group in the end. Holding back and not sharing information forces people apart, while sharing knowledge and resources brings them together. It closes the divide or gap.

I am very grateful that AI models were released when they were. They helped me immensely during this time as a resource to aid in finding, understanding, and processing information. Without them, I wouldn't have gotten this far.

Advancements in technology and AI are necessary for us to continue progressing. We can learn and develop at faster rates than ever before. But sometimes, when growth happens too quickly, it can lead to issues and unforeseen problems. Even social media algorithms can be used to feed information and manipulate. We need to ensure these technologies are used for good, not evil. Currently, we lack the proper regulations and legal protections to keep individuals and our societies safe.

Like with Evan Seyfried and the hypothetical scenario of forcing someone off a cliff without actually touching them, neurotechnology can be used to drive someone off this hypothetical cliff. Until this technology is properly exposed in the mainstream, it will continue to be a threat to all of humanity. Without the proper exposure and acknowledgment of its capabilities, we will never have the proper means to protect ourselves against it.

Corporations, government agencies, medical professionals, law enforcement, and judicial systems can all act with negligence because of the lack of exposure to the truth about these technologies. People will continue to suffer due to negligence and fear of what proper exposure could do to the economy, for example. How much is someone's life worth?

To regulate, we first need to understand the problem at hand. What are these new technologies capable of, and what are their effects? How far can they go, and to what extent can they cause damage and harm? But before we can even ask these questions, we must first admit that the technology exists so we can address the issues. Cognitive liberties are at risk as technology increasingly interfaces with human brain function. It is apparent that this has already gone too far and reached a tipping point. It has to stop, and things need to change on a global level. It is a matter of freedom of thought and free will.

11

FOR THE FUTURE

EFFICACY: SELF, COLLECTIVE, AND SUPPORTIVE

Efficacy is a term referring to the ability or power of something to produce a desired outcome. Regarding AI, for example, the more appropriately it is trained with quality data for a particular application, the greater its efficacy will be. With regard to humans, I consider three separate categories: self-efficacy, collective efficacy, and supportive efficacy.

Self-efficacy is similar to self-confidence but is more specifically geared toward completing a task. Someone with high self-efficacy strongly believes in their ability to successfully complete a task. They draw information from past examples and can work to fill in the gaps as needed. With advancements in AI, we can all be more successful at completing most tasks until one day AI will be completing most tasks for us. My self-efficacy helped me continue through this time. I may be shy, but I do not shy away from complex problems; if anything, I am driven by them.

Collective efficacy is a similar concept but applies to a group's shared belief in their ability to succeed in a given task. This is important for the effectiveness of any team. I believe it is extremely important for humanity as a whole. There is also potential for AI to take over in the future, and we have to take action now to ensure that does not happen.

The third term is supportive efficacy. I have not found this to be a formally defined term, but I strongly believe it should be, so I will define it here. Supportive efficacy is when one individual believes in another individual's ability to successfully complete a task. It's encapsulated in that very simple yet powerful phrase: "You got this." People should be able to choose what they want to do in life and be supported in those choices with respect and dignity. If we had more of this in our workplaces and relationships instead of discrimination, harassment, and degradation, our society would be much better off.

HEALING

There is no question about it, being put through this type of experience will cause pain and harm. But a gift we have as humans is the ability to heal. Whether emotional or physical, we are capable of healing. Believe in that.

Emotional pain and grief can take time to process and reach acceptance. However, we are able not only to heal, but to come back stronger and wiser. I can't deny that I have learned a great deal of information through this. They are lessons learned and lessons that can be shared.

The most concerning part has been the physical impacts on my brain. Brain neuroplasticity thankfully allows for healing of the brain. What seemed to help me towards recovery the most, after the attacks finally stopped, was restful sleep and oil supplements along with exercise and a healthy diet.

Maybe the intensions all along were to knock me out of my career and future success. However, they only made me what I am today. I refuse to let this be the end of me. This is only the beginning.

ETHICS AND AI

Ethics is crucial in building healthy work environments and a sustainable future for humanity. Maintaining ethics and good faith is essential in fostering trust and creating unity. Each individual is responsible for upholding their moral principles to ensure ethical decision-making. What may be even more difficult is instilling this type of ethical standard into AI.

In terms of employment law, most of the U.S. operates under at-will employment. At-will employment means an employer can terminate employment at any time without reason, as long as the grounds for such termination are not illegal. Three potential exceptions to at-will employment are Public Policy, Covenant of Good Faith and Fair Dealing, and Implied Contract. The Public Policy exception protects employees when they exercise rights under public policies, such as serving jury duty. The Covenant of Good Faith and Fair Dealing requires both employers and employees to act honestly and fairly, while the Implied Contract exception allows conditions of employment to be based on statements made in employee handbooks, among other sources. Not all states apply each of these exceptions, but those that do can be safer for employees.

The achievement of ethical AI depends on parameters like fairness, transparency, and privacy. Is AI acting with fairness, or is it biased in some way, whether through ignorance or prejudice? Do we have full transparency regarding AI and advanced technology? Not yet. And is our privacy and data secured adequately, with appropriate rights and laws as well as in practice? Definitely not yet.

PROTECTIONS AND CONSCIOUSNESS

Reflecting again on my own experiences, I believe that being aware of your rights is most vital. To make appropriate complaints and avoid gaslighting and negligence, one must first understand the laws and how to protect oneself. This makes speaking up and preventing further wrongdoing much more attainable. After all, how can you use the law to protect yourself if you are unaware of it?

The circumstances I went through lasted years. Individuals and entities repeatedly worked against me to prevent my success; however, there were instances where the neurotechnology protected me by providing information. The difficulties lie in the lack of availability or means to monitor and track neurotechnology.

This technology and AI may be close to obtaining digital superintelligence. What has already been accomplished is the monitoring and retention of human consciousness. The fact that this is not yet regulated is alarming. Simultaneously, the thought of what the future could hold is exciting. This data and technology could provide protections instead of threats and vastly improve the quality of health, safety, and life. A bright future is never guaranteed, but I believe in us.

APPENDIX

THE DOCTRINE OF BLESSEDNESS

This doctrine helped me through this difficult time.
Jesus Christ's Beatitudes:

1. Blessed are the poor in spirit,
   for theirs is the Kingdom of Heaven.

2. Blessed are those who mourn,
   for they will be comforted.

3. Blessed are the meek,
   for they will inherit the Earth.

4. Blessed are those who hunger and thirst
   for righteousness,
   for they will be satisfied.

5. Blessed are the merciful,
   for they will be shown mercy.

6. Blessed are the pure in heart,
   for they will see God.

7. Blessed are the peacemakers,
   for they will be called the children of God.

8. Blessed are those who are persecuted
   because of righteousness' sake,
   for theirs is the Kingdom of Heaven.

9. Blessed are ye, when men shall revile you, and
   persecute you, and shall say all manner of evil against
   you falsely, for my sake. Rejoice, and be exceeding
   glad, for great is your reward in heaven.

BEATITUDES INTERPRETED

The following are my interpretations of the implications of these beatitudes, specifically after this experience:

1. Do not chase after money.

2. Do not be afraid to get emotional.

3. Do not seek revenge.

4. Do try to do the right thing.

5. No one is perfect.

6. Be good and kind.

7. Choose peace over conflict.

8. Prove the truth even when everyone is against you.

9. Do not worry about the false narratives and keep going.

www.ingramcontent.com/pod-product-compliance
Lightning Source LLC
Chambersburg PA
CBHW022046210326
41519CB00055B/846